Personal Computing

IBM PC Upgraders Manual
Revised Edition
1990 0 471 52452 2

IBM PS/2 User's Reference Manual
1990 0 471 62150 1

Communicating With the IBM PC Series
1988 0 471 91667 6

DOS Productivity Tips & Tricks
1988 0 471 60895 5

The Laptop/ Notebook Companion
1992 0 471 55804 4

Reference

The Complete Modem Reference
1991 0 471 52911 7

The Complete PC AT and Compatibles Reference Manual
1991 0 471 53315 7

Data and Computer Communications: Terms, Definitions and Abbreviations
1989 0 471 92066 5

THE MULTIPLEXER REFERENCE MANUAL

THE MULTIPLEXER REFERENCE MANUAL

Gilbert Held
4-Degree Consulting
Macon, Georgia, USA

JOHN WILEY & SONS
Chichester · New York · Brisbane · Toronto · Singapore

Other Wiley Editorial Offices

John Wiley & Sons, Inc., 605 Third Avenue,
New York, NY 10158-0012, USA

Jacaranda Wiley Ltd, G.P.O. Box 859, Brisbane,
Queensland 4001, Australia

John Wiley & Sons (Canada) Ltd, 22 Worcester Road,
Rexdale, Ontario M9W 1L1, Canada

John Wiley & Sons (SEA) Pte Ltd, 37 Jalan Pemimpin #05-04,
Block B, Union Industrial Building, Singapore 2057

Library of Congress Cataloging-in-Publication Data:

Held, Gilbert, 1943–
The multiplexer reference manual / Gilbert Held.
p. cm.
Includes index.
ISBN 0 471 93484 4
1. Multiplexing. I. Title.
TK5102.5.H453 1992
621.39′81—dc20 92-19944
CIP

British Library Cataloguing in Publication Data:

A catalogue record for this book is available from the British Library

ISBN 0 471 93484 4

Typeset in 11/13pt Bookman from author's disks by Text Processing Department,
John Wiley & Sons Ltd, Chichester
Printed and bound in Great Britain

CONTENTS

Preface xi

1 Introduction 1

1.1 Rationale for Multiplexing 2
Economics 2
Diagnostic testing 5
Error detection and correction 6
Network management support 6
Switching and routing 7
Voice/data integration 9
1.2 Equipment Evolution and Utilization 10
Frequency division multiplexers 10
Time division multiplexers 12
Statistical multiplexers 14
Packet assemblers/disassemblers 15
T-carrier multiplexers 17
Fiber optic multiplexers 18
1.3 Future Developments 19

2 Frequency Division Multiplexers 21

2.1 Types of Frequency Division Multiplexing 21
2.2 Communications Carrier Systems 22
CCITT FDM wideband recommendations 23
The standard group 23
The standard supergroup 24
The standard mastergroup 25
2.3 FDM Data Multiplexers 26
Carrier telegraph 26
Modern FDM data multiplexers 28
Data multiplexing capability 30
FDM utilization 33

3 Time Division Multiplexing 37

3.1 FDM versus TDM 37

3.2 TDM Operation 38
I/O channel adapter 38
Central logic 40
Composite adapter 40
The message frame 41
The multiplexing interval 42
TDM techniques 43
TDM efficiency 45
3.3 TDM Applications 46
Point-to-point multiplexing 47
Series multipoint multiplexing 48
Hub-bypass multiplexing 49
Front-end substitution 50
Inverse multiplexing 51
3.4 Multiplexing Economies 52
Networking example 53
Alternative configurations 56
Combined FDM–TDM 58
3.5 Other Types of TDMs 61
Multiport modems 61
Operation 61
Selection criteria 62
Application example 63
Standard and optional features 66
Split stream DSUs 68
Applications 69

4 Statistical Multiplexers 71

4.1 Comparison to TDMs 71
TDM message frame 72
Statistical frame construction 74
Address and byte count method 75
Dual addressing format 76
Data integrity 78
Operational problems 78
4.2 Buffer Control 79
Inband and outband signaling 79
Clocking signal adjustment 79
Flow control variations 80
Buffer delay 81
4.3 Service Ratios 83
Examples 84
Data source support 85
ITDMs 86
4.4 STDM Statistics 87
Multiplexer loading 88
Buffer utilization 88
Buffer occupancy 88
Frame transmitted and negative acknowledgements 88
Ratios 89
Compression efficiency 89

Statistical loading 89
Character error rate 90
4.5 STDM Features 90
Auto speed and auto code detection 90
Auto echo 91
Bandpass 92
Command console 92
Data compression 92
Flyback delay 92
Protocol support 93
Port contention 93
Multi-node 93
Statistics display 94
4.6 STDM Applications 94
Multi-node networking 94
Building STDM networks 97

5 Packet Assembler/Disassembler 99

5.1 Evolution 99
5.2 CCITT X Series Recommendations 100
X.25 100
Physical level 101
Data or link level 101
Packet level 101
X.3 102
X.28 102
X.29 102
5.3 PAD Operation 104
Packet assembly 104
Packet framing 104
5.4 PAD States 105
Command state 105
Data transfer state 106
5.5 Access Request Operation 106
BT Tymnet 106
Sprint Net (Telenet) 108
5.6 X.3 Parameters 108
1:m PAR recall 110
2:m Echo 110
3:m Selection of data forwarding signal 110
4:m Selection of idle timer delay 110
5:m Ancillary device control 110
6:m Control of PAD service signals 111
7:m Procedure on receipt of break signal 111
8:m Discard output 111
9:m Padding after carriage return 111
10:m Line folding 111
11:m Binary speed 112
12:m Flow control of the terminal PAD 112
13:m Line feed insertion after carriage return 112
14:m Padding after line feed 112
15:m Editing 113

16:m Character delete 113
17:m Line delete 113
18:m Line display 113
20:m Echo mask 113
21:m Parity treatment 114
22:m Page wait 114
5.7 X.28 Commands 114

6 T-Carrier Multiplexers 117

6.1 T1 Circuit Evolution and Operation 117
Channel banks 118
Channel banks versus T-carrier multiplexers 120
Framing structure overview 120
Signaling restrictions 123
6.2 T-carrier Multiplexers 125
Operational characteristics 125
CSU function 125
Application overview 127
Multiplexing efficiency 128
6.3 Features to Consider 129
Bandwidth utilization 129
Bandwidth allocation 131
Voice interface support 133
Loop signaling 133
E&M signaling 134
Voice digitization support 135
PCM 135
ADPCM 135
CVSDM 135
Internodal trunk support 138
Subrate channel utilization 139
Digital access cross connect capability 139
Gateway operation support 141
Alternate routing and route generation 142
Redundancy 143
Maximum number of hops and nodes supported 144
Diagnostics 144
Configuration rules 144
Multiplexers and nodal processors 145
6.4 Networking 146
Point-to-point single link network 146
Point-to-point multiple link network 146
Star network 147
Ring network 149
Multipoint network 149
Mesh network 151

7 Fiber Optic Multiplexers 153

7.1 System Components 154
The light source 154
Optical cables 155

Types of fibers 157
Cable size 157
Common cable types 157
The light detector 159
7.2 The Optical Modem 159
7.3 Optical Transmission Advantages and Limitations 161
Bandwidth 161
Electromagnetic non-susceptibility 162
Signal attenuation 163
Electrical hazard 163
Security 163
Weight and size 164
Durability 164
Limitations of use 164
Cable splicing 165
System cost 165
7.4 Utilization Economics 166
Dedicated cable system 166
Multichannel cable 167
Optical multiplexers 167
7.5 Types of Fiber Optic Multiplexers 169
Selection considerations 169
Physical interface 171
Fiber optic modem 172
7.6 SONET and SONET Multiplexers 172
Frame structure 173
Possible applications 174

8 Evolving Technologies 177

8.1 Low Bit-rate Voice Digitization 177
Advantages 177
Applications 178
8.2 Fast Packet Multiplexing 179
Operation 179
Applications 180
Standards 180
8.3 Frame Relay 181
Protocol support 181
Operation 182
8.4 Alternatives to Consider 182

Index 185

PREFACE

The multiplexer can be considered as one of the most important types of communications products developed during this century. Originally designed as a mechanism to reduce the cost of communications by the application of frequency division, both the method of operation and functions performed by multiplexers have significantly changed.

Through the use of microprocessor technology, a high degree of intelligence has been incorporated into many multiplexing products. This in turn has resulted in the multiplexer evolving into a key networking component used by most commercial organizations, government agencies, and public and private universities. Today, multiplexers form the basis for developing intelligent networks, integrating voice and data onto common high-speed transmission lines, and providing access to multiple computers via a common network, among other functions.

As we move into the next century multiplexers can be expected to continue their leading role in providing users with the capability to develop intelligent networks. Although we can expect additional features, functions, and capabilities to be added to this communications device, its primary function to provide an economical method to interconnect data sources at dispersed locations will continue to provide the primary rationale for using multiplexers.

Based upon a series of lectures and seminars I conducted covering different data communications topics in the United States, Europe, and Israel, I received many requests for additional information concerning the operation and utilization of multiplexers. Although I have previously incorporated a limited amount of information concerning multiplexers into several of my books, I decided the level of interest of persons I meet, as well as the importance of the subject, justified a book devoted to multiplexers. In this book I have attempted to place into one volume comprehensive information concerning the different types of multiplexers that network analysts, designers, operators, and practitioners can consider; how they operate, their features, and most important, numerous examples of network applications involving the use of multiplexers. From reading this book, both lay and experienced network personnel should obtain an additional insight into the operation and utilization of different types of multiplexers. As always, I welcome reader comments that can be sent directly to me or through my publisher.

1

INTRODUCTION

Unlike many communications terms, the word 'multiplex' as well as the equipment designator 'multiplexer' can be found in many dictionaries. In examining the meaning assigned to these terms you will note multiplex being denoted as the process of 'simultaneously transmitting several messages,' while the term multiplexer will refer to a device which 'provides the capability to simultaneously transmit several messages.' Although both definitions are essentially correct, like many communications books they simply define a major function without considering the usefulness of the process and the tremendous capabilities of equipment designed to perform the process of multiplexing. Thus, the goal of this book is to fill a void with respect to the operation and utilization of different types of multiplexers.

As an introductory chapter, the focus of attention is upon the rationale for multiplexing and the evolution of equipment that performs multiplexing. By examining the rationale for multiplexing, we will obtain a detailed overview of the functionality and capability of multiplexers in addition to understanding why and when we should consider their use. Although this book was certainly not written as an historical novel, understanding the evolution of multiplexers will also provide us with significant, useful information. From an understanding of the evolution of multiplexers, we can obtain an insight into their technological evolution. This insight will provide us with knowledge concerning both their present and expected utilization, with the latter allowing the network analyst, designer, and practitioner to properly plan for using tomorrow's technology today.

1.1 RATIONALE FOR MULTIPLEXING

Approximately 20 years ago when I was introduced to multiplexer technology in a classroom environment, the primary and probably only reason to use multiplexers was one of economics. Then, as today, the statement 'whenever data lines run in parallel, consider multiplexing' holds true. However, the rationale for using multiplexers while still having economics as a major driving force has considerably expanded. For example, Table 1.1 lists five of the more popular reasons that form the basis for onsidering the use of different types of multiplexers, in addition to economics, which heads the list. To provide readers with a firm understanding of the entries in Table 1.1, let us examine each. In doing so we will consider one or more examples of why each entry can form the basis for a decision to use multiplexers.

Table 1.1 Rationale for using multiplexers

- Economics
- Diagnostic testing
- Error detection and correction
- Network management support
- Switching and routing
- Voice/data Integration

Economics

As previously discussed, the old adage 'when lines run in parallel, consider multiplexing' holds true today. To illustrate why economics plays an important role in a decision to multiplex, we can examine the routing of parallel communications lines as illustrated in Figure 1.1A. In this example two lines are shown routed between common locations designated as X and Y.

If we assume the monthly cost of each communications line is k dollars, pounds, or other financial units, then the total monthly cost is $2k$. To illustrate why economics forms the primary rationale for the use of multiplexers, consider Figure 1.1B in which the parallel communications lines routed between locations X and Y were replaced by the use of a single line and a pair of multiplexers, each denoted by a triangular symbol. If the monthly cost of leasing each multiplexer is m, then whenever

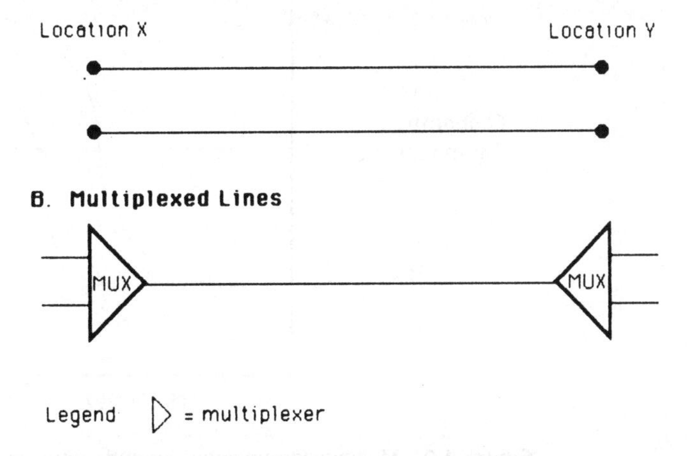

Figure 1.1 Multiplexing economics. The replacement of two or more lines by one line through the use of multiplexers (MUX) forms the basis for economic savings provided by the use of multiplexers.

$2m + k < 2k$, it is more economical to use multiplexers than run a pair of communications lines in parallel.

Since the monthly cost of a communications line is distance sensitive, with longer distance lines costing more than shorter distance lines, we can generalize the economics associated with multiplexing based upon the number of lines to be multiplexed and the distance between multiplexing locations. Figure 1.2 illustrates the potential economic savings associated with the use of multiplexers based upon the number of lines multiplexed and the multiplexing distance. In examining Figure 1.2, let us first discuss the positive economic savings and then focus our attention upon the reason why multiplexing is not always economical, denoted by the negative economic savings portion of Figure 1.2.

In examining the positive economic savings portion of the curves illustrated in Figure 1.2, we can use the previously indicated inequality $2m + k < 2k$ to denote the basis for the construction of each curve. First, as the multiplexing distance increases, the cost of each line increases, since cost is proportional to distance. Since the cost of a pair of multiplexers is fixed while the line costs denoted by k are variable, as the multiplexing distance increases, k increases, resulting in an increased level of economic savings beyond the breakeven point. Now suppose an additional circuit is multiplexed. Then,

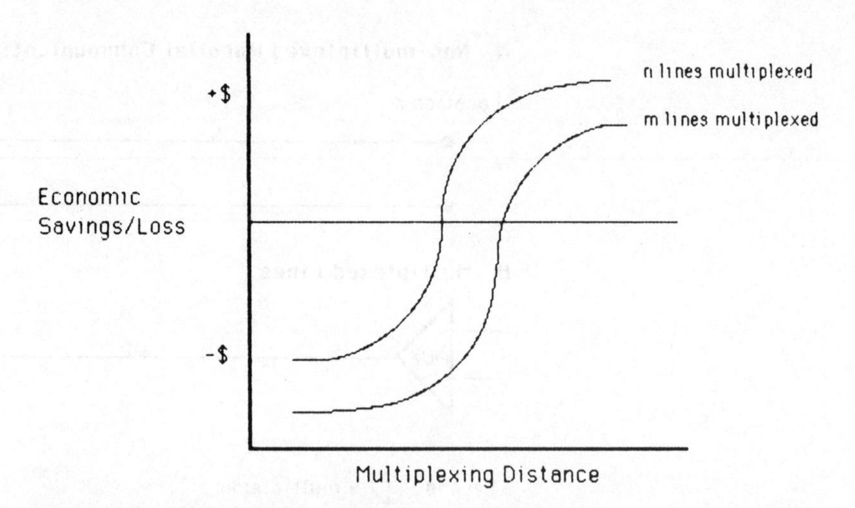

Figure 1.2 Multiplexing savings depends upon the number of lines multiplexed and the distance between multiplexed locations. From an economic perspective, multiplexing becomes more effective as the number of lines between multiplexing locations and the distance between locations increases.

for multiplexing to be economical the inequality $2(m + \Delta m) + k < 3k$, where Δm represents the incremental cost in upgrading a multiplexer to multiplex an additional line. The incremental cost associated with upgrading a multiplexer to support an additional data source is normally a fraction of the cost associated with another separate line. Thus, the economic savings associated with multiplexing increases as the number of lines to be multiplexed increases. This explains the reason why the curve labeled 'n lines multiplexed' rises faster and has a greater y axis value than the curve labeled 'm lines multiplexed', since it is assumed that $n > m$.

Although the use of multiplexers can result in economic savings, this situation is not guaranteed. In fact, since the lease or purchase of multiplexers represents either a repeating monthly lease expense or a large one-time expenditure, line savings must exceed the lease or over a period of time exceed the one-time expenditure for multiplexing to be economically effective. For relatively short line distances, the cost savings associated with replacing several lines by one typically will not exceed the cost of the multiplexing equipment, resulting in a negative economic savings level. This is illustrated by the portions of the two curves that decrease in tandem with a decrease in multiplexing distance to the left of their breakeven points.

Diagnostic testing

Most multiplexers include a built-in diagnostic testing capability. This testing capability can range in scope from a self-test of the device, in which a predefined test message is looped through the equipment, to sophisticated tests that can be used to ascertain the status of a remote multiplexer as well as the quality of the circuit interconnecting multiplexers.

Regardless of the type of diagnostic testing supported, this functional capability can be a valuable asset that can assist you in isolating and correcting networking problems. For example, consider the small network schematic diagram illustrated in Figure 1.3 in which several terminal users are connected via a multiplexer to a distant computer center. Suppose several terminal users report the inability to communicate with the mainframe. Without further information, the cause of the problem could be the local or distant multiplexer, the circuit connecting the multiplexers, or the mainframe computer.

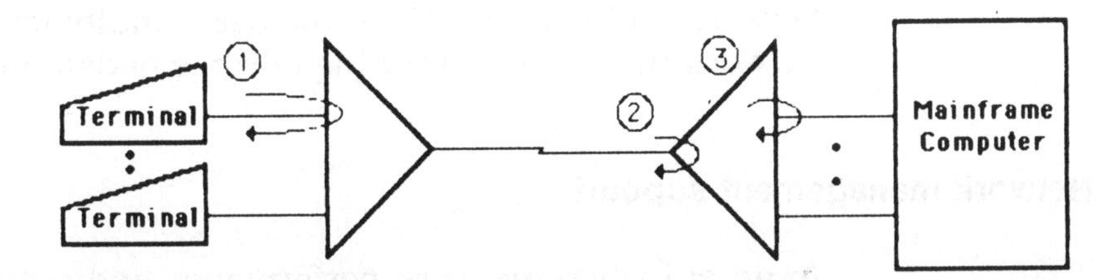

Figure 1.3 Multiplexer diagnostics can assist in isolating communications problems. Through multiplexer loopbacks, you can check (1) the cabling of a terminal to a multiplexer port, (2) the operation of the local multiplexer and circuit, or (3) the operation of both multiplexers and the circuit connecting the multiplexers.

Through the use of the diagnostic testing capability built into most multiplexers, you should be able to determine the operational state of each device. If each device passes its self-test and supports remote testing, you could next loop the circuit through the distant multiplexer and initiate another diagnostic test to determine if the circuit is causing the reported problem. Note that you can perform these tests and more than likely additional tests without the use of test equipment, thus providing you with the ability to economically isolate many network problems and initiate action to correct such problems.

Error detection and correction

Some types of multiplexers provide essentially end-to-end error-free transmission. To provide this capability such multiplexers block data and apply an algorithm to the block using a predefined mathematical process. This process results in the generation of one or more 'check' characters that are added to the block. The receiving multiplexer applies the same algorithm to the received data block, using the same predefined mathematical process. This results in the local generation of one or more 'check' characters which are compared to those characters added to the transmitted block. If the check character or characters match, the data block is assumed to have been received correctly, otherwise one or more bit errors are assumed to have occurred. To correct the erroneously received block the receiving multiplexer will request the sending multiplexer to retransmit the block. Hence, errors are corrected by retransmission.

The error detection and correction capability of many multiplexers results in essentially error-free transmission between multiplexers. Thus, the use of multiplexers can provide network users with a very high degree of data integrity.

Network management support

In an era where network performance and availability are key operational considerations, the ability of most multiplexers to operate under the control of a network management system is a most important consideration. Doing so can provide technical control center personnel, network analysts, and communications managers, among others, with the ability to obtain valuable information concerning the state of data flow through an organization's network. This in turn may provide an astute observer with information that can be used to prevent network bottlenecks prior to their occurrence. For example, from the statistics provided by multiplexers you may observe over a period of time an increase in the utilization of a circuit connecting two multiplexers. This might suggest the use of a higher-capacity circuit and an upgrading of the multiplexers to prevent a potential bottleneck from occurring.

A second area where the integration of multiplexers into a network management system can provide significant benefits is in the automatic rerouting of information when circuits fail or equipment becomes inoperative. For example, consider the three-location network illustrated in Figure 1.4. Through the use of a network management system supported by the multiplexers you may be able to configure each multiplexer to initiate a predefined rerouting of data whenever a circuit fails. Thus, the failure of the circuit interconnecting locations B and C could be automatically compensated for by the rerouting of data from location B to location C via location A. Although this example may appear simplistic, suppose your organization's network had 25, 50, or perhaps even 100 or more locations interconnected by a variety of circuits. By being able to preplan alternative courses of action and configure your equipment to implement those courses of action based upon the occurrence of predefined events, you can ensure that your equipment would automatically adjust to most, if not all, exigencies. This in turn can increase the level of network performance and availability.

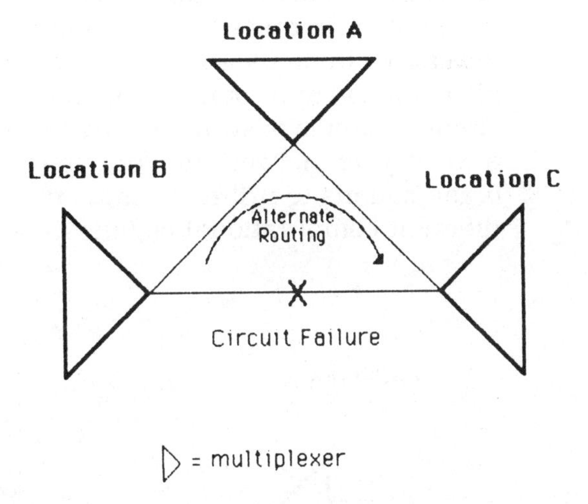

Figure 1.4 Alternate routing. Alternate routing provides the ability to compensate for circuit failures and equipment malfunctions.

Switching and routing

In our previous discussion of network management support we examined a simple example of multiplexer routing. A related, but not always identical, capability to routing is switching.

Most multiplexers support one or only a few output ports that carry multiplexed traffic. Thus, many times when we refer to the routing capability of a multiplexer we mean the ability of the multiplexer to take the multiplexed transmission destined for one port and place that traffic on a different port. Doing so provides an alternate route between nodes in a network as previously illustrated in Figure 1.4.

In comparison to routing, which we could refer to as switching of multiplexed traffic, switching involves the transfer of demultiplexed data to specific ports or port groups. Concerning the latter, port groups may be configured to operate under multiplexer switching similar to a telephone rotary or hunt group. That is, if the first port in the group is occupied, the multiplexer will then attempt to switch to the next port in the group.

Figure 1.5 illustrates a simple example of multiplexer switching. In this example, the multiplexer at location B functions as a switching multiplexer and was configured so that its output ports are connected to two different computers. Each group of ports routed to a computer functions as a group and, depending upon the switching capability of the multiplexer, may also function as a rotary group. Although Figure 1.5 illustrates a simple example of switching, this feature can be used to provide a variety of networking functions, including those enabling organizations to utilize a common network to provide access to different computational facilities.

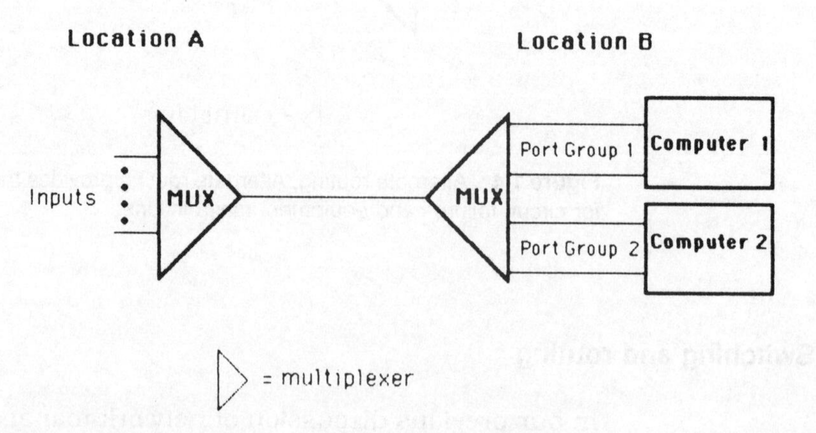

Figure 1.5 Multiplexer switching. Multiplexer switching involves the routing of demultiplexed traffic to specific ports or to port groups.

When a network is designed based upon the use of switching multiplexers, virtual circuits are initiated by each switching multiplexer. The resulting network is commonly referred to as a circuit switched network and should be distinguished from a packet switched network. Here the former type of network results in the exclusive use of the established switched circuit by the data source being switched. Concerning the latter type of network, data is grouped into units called packets, and many data sources represented by individual packets can be routed over a common circuit. Similar to the other features listed in Table 1.1, we will examine multiplexer switching and routing in more detail later in this book. In addition, we will examine the operation and utilization of specialized equipment that forms packets, since such equipment can be considered to be a special type of multiplexer.

Voice/data integration

The commercial availability of the T1 circuit in the mid-1980s in North America and the E1 circuit in Europe resulted in the development of the T1/E1 multiplexer, the latter providing numerous organizations with a platform to integrate their voice and data transmission requirements. This in turn allowed the merging of previously separate voice and data networks, enabling the commonality of facility usage not only to provide economic savings, but, in addition, to facilitate network management control which could provide common backup and alternate routing for both voice and data network users.

One reason for the widespread acceptance of T1/E1 multiplexers was the development of voice digitization techniques. Through the incorporation of such techniques into their multiplexers, vendors could support from double to quadruple the number of simultaneous voice conversations capable of being carried on a T1/E1 circuit. Within a short period of time multiplexer vendors began to provide a similar voice digitization capability for their products designed to communicate over lower-speed digital transmission facilities. Today, you can obtain a wide range of multiplexers that provide you with the capability to integrate your organization's voice and data communications requirements. In doing so multiplexers provide you with the platform required to implement a common voice/data network.

1.2 EQUIPMENT EVOLUTION AND UTILIZATION

From an historical perspective, multiplexing technology can trace its origins to the early development of telephone networks. Then, as today, multiplexing was the employment of appropriate technology to permit a communications circuit to carry more than one signal at a time.

The earliest known application of multiplexing occurred in 1902, 26 years after the world's first successful telephone conversation. In an attempt to overcome the existing ratio of one channel to one circuit, telephone companies used specially developed electrical network terminations to derive three channels from two circuits. The third channel was denoted as the phantom channel; hence the name 'phantom' was applied to this early version of multiplexing. Although this technology permitted two pairs of wires to effectively carry the load of three, the requirement to keep the electrical network finely balanced to prevent crosstalk limited its practicality.

Frequency division multiplexers

The first modern application of multiplexing technology involved the subdivision of the frequency bandwidth of lines routed between telephone company central offices into smaller segments called data bands or derived channels. This technique which is illustrated in Figure 1.6 is known as frequency division multiplexing (FDM) and results in each data band being separated from another data band by a guard channel. Here the guard channel prevents a frequency drift occurring on one band from interfering with data carried on a second band. By using a wideband line capable of carrying a large range of frequencies between central offices and subdividing that line by frequency, many voice conversations could be simultaneously carried between central offices. As calls were routed through a central office, they were shifted in frequency to occupy a vacant data band on the wideband line. Then, at the destination central office, the call was shifted back to its original frequency and passed to the subscriber connected to that central office.

The growth in the number of telephone company subscribers through the 1960s was accompanied by an increase in the number of long-distance calls. This in turn resulted in an increase in the use of FDM to economize upon the number of lines required to interconnect telephone company central

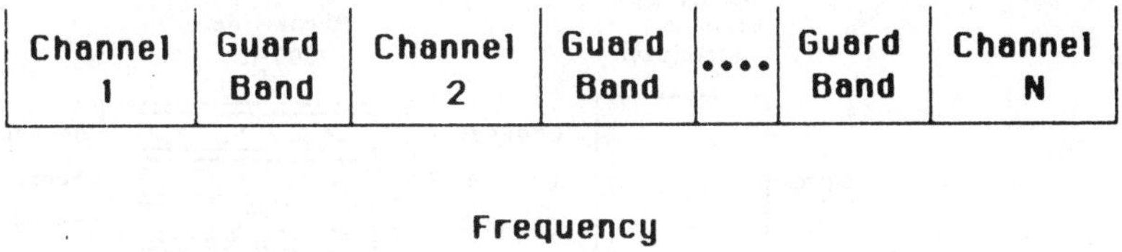

Channel 1	Guard Band	Channel 2	Guard Band	....	Guard Band	Channel N

Frequency

Figure 1.6 Frequency division multiplexing. In frequency division multiplexing, the bandwidth of the circuit is subdivided into derived channels or data bands, separated from each other by guard bands.

offices. The 1960s was also notable for the establishment of computer networks by banks, insurance companies, and other commercial organizations. To assist those companies as well as government agencies in establishing economical networks, several manufacturers applied frequency division multiplexing techniques to voice-grade leased lines, resulting in the development of equipment that could be used to simultaneously carry several 'data conversations.'

In developing FDM for data applications, the 3000 Hz bandwidth of an analog leased line was subdivided into subchannels or derived bands similar to the method by which telephone companies subdivided their wideband lines. However, instead of shifting a voice conversation by frequency, the FDM developed to carry data conversations places tones on the line in a predefined subchannel to correspond to the bit composition of the data. That is, a tone is placed on the line in a specific subchannel at one frequency to correspond to a binary one, while a second tone at another frequency is placed on the line to correspond to a binary zero. This modulation technique is known as frequency shift keying (FSK) and was borrowed from FSK modems developed during the 1960s.

Figure 1.7 illustrates the operation of a four-channel FDM designed to carry four data conversations. Since each digital data stream is converted into a series of tones at predefined frequencies, the modulation of data is performed by the multiplexer, eliminating the necessity of having modems to perform this function. To prevent the tones on one channel from drifting into another channel, this type of FDM also employs the use of guard bands.

Although FDM is still in use, its peak period of implementation was between the late 1950s and mid-1960s, after which it began to be replaced by time division multiplexing (TDM) technology. This newer technology permitted an increase in

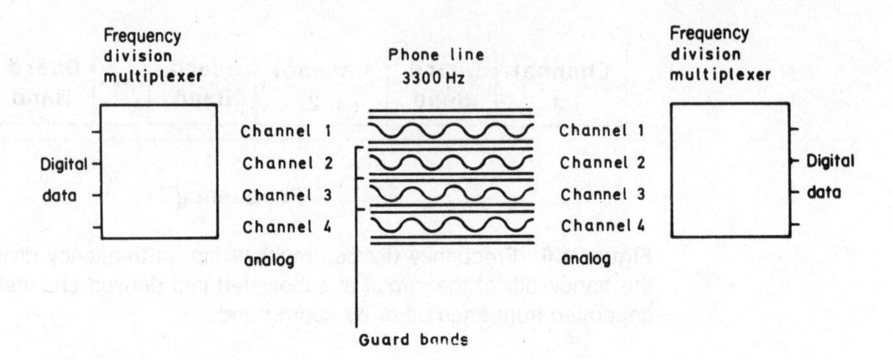

Figure 1.7 FDM used for data.

line utilization which could result in a greater capability for economic savings. Another advantage of TDM is the fact that it is based upon digital technology which, in general, is significantly more reliable than analog-based FDM equipment. In spite of the superior performance and reliability of TDM over FDM, a few organizations and many communications carriers still maintain FDM equipment, and its utilization can be expected to continue on a limited basis through the beginning of the next century.

Time division multiplexers

The time division multiplexer (TDM), as its name implies, separates the high-speed composite line into intervals of time. The TDM assigns a time slot to each low-speed channel whether or not a data source connected to the channel is active. If the data source is active at a particular point in time the TDM places either a bit or a character into the time slot, depending upon whether the TDM performs bit interleaving or character interleaving multiplexing. If the data source is not active the TDM places an idle bit or character into the time slot. At the opposite end of the high-speed data link another multiplexer reconstructs the data for each channel based upon its position in the sequence of time slots. This process is called demultiplexing.

The top of Figure 1.8 illustrates the use of TDMs to multiplex four data sources onto one high-speed line. The lower portion of Figure 1.8 illustrates how the TDMs interleave bits or characters into time slots on the high-speed line at one end of the communications link (multiplexing) and remove the bits or

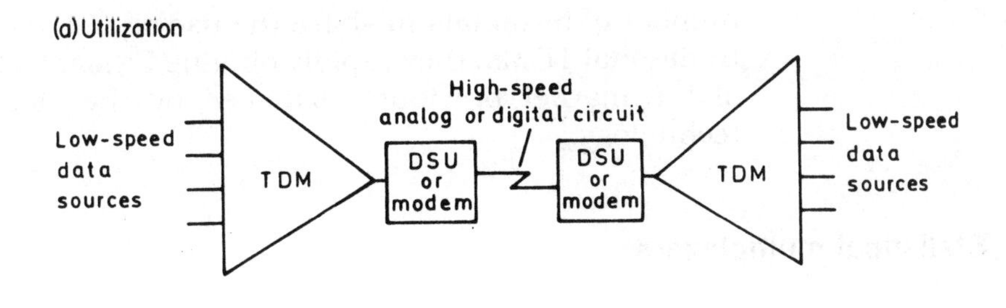

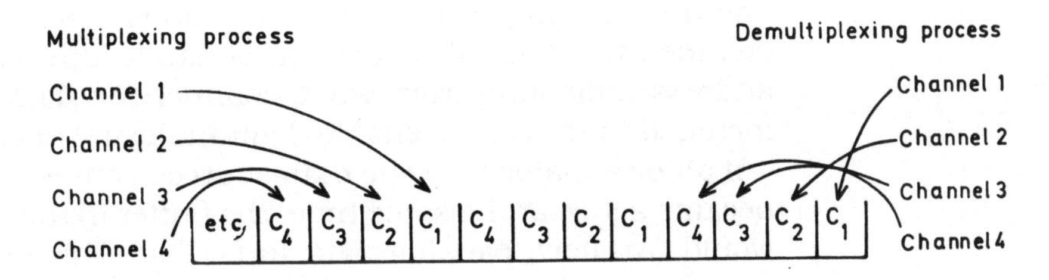

Figure 1.8 Utilization (top) and operation (bottom) of time division multiplexing.

characters from the high-speed line (demultiplexing), sending them to their appropriate channels.

Today, the most commonly used technique for multiplexing data involves the subdivision of a circuit into time slots into which data sources are placed. This technique includes both time division multiplexing in which data sources are assigned to fixed time slots and statistical multiplexing in which the assignment of data sources to time slots is variable, based upon numerous factors, including the operation of the multiplexer and the presence or absence of data on each channel being multiplexed.

TDM equipment was originally based upon fixed logic circuits. With the incorporation of microprocessor technology into multiplexers, the intelligence and programmability of the microprocessor enabled the dynamic allocation of time slots. Since this allocation process was statistical in nature, based upon the activity of data sources connected to other multiplexer channels, the device was called a statistical multiplexer or STDM.

Statistical multiplexers reached the commercial marketplace in the early 1970s. Due to their capability to enable a larger number of terminals to share the use of a high-speed line than traditional TDMs, they rapidly obtained a significant percentage of the market previously satisfied by the older multiplexing technology.

Statistical multiplexers

The key to the efficiency of a statistical multiplexer is its dynamic allocation of bandwidth. In this technique a microprocessor in the multiplexer only places data into a time slot when a device connected to the multiplexer is both active and transmitting. Since a time slot can now hold data from any channel, the dynamic allocation of bandwidth requires that an address indicating from what channel the data originated be included in the slot to enable demultiplexing to occur correctly.

If all or a majority of the data sources connected to an STDM became active at the same time, the buffer in these multiplexers would overflow, causing data to be lost. This situation would occur since the composite data rate of active channels would exceed the operating rate of the high-speed line, causing the buffers in the multiplexers to fill and eventually overflow. To prevent this situation from occurring, STDMs incorporate one or more techniques that inhibit data transmission into the multiplexer when the data in its buffer reaches a predefined level. Then, after data is transferred from the buffer onto the composite high-speed line the buffer occupancy is lowered until another predefined level is reached. This lower level then becomes a trigger mechanism for the multiplexer to enable previously inhibited data sources to resume transmission.

The process of inhibiting and enabling data transmission is known as flow control. The most common method of flow control is obtained by lowering and raising the clear to send (CTS) control signal. Another common method of flow control is obtained by having the multiplexer transmit the XOFF and XON characters to devices that recognize those characters as a signal to stop and resume transmission. Figure 1.9 illustrates the statistical multiplexing process.

Since the late 1970s, an evolution in communications technology and the growth in the usage of packet networks formed the basis for the development and introduction of several specialized types of multiplexers. Although these multiplexers are based upon TDM and STDM technology, what separates them from other multiplexers is the fact that their usage is

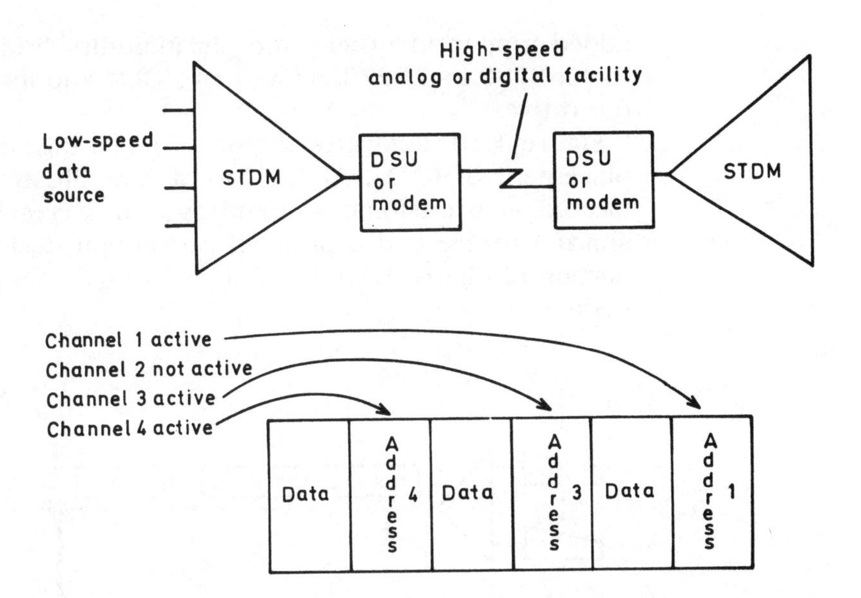

Figure 1.9 Utilization (top) and operation (bottom) of statistical multiplexing. Statistical multiplexers dynamically allocate data to time slots when data sources are active.

restricted to specific communications markets. Multiplexers in this category include packet assemblers/disassemblers, T-carrier multiplexers, and fiber optic multiplexers.

Packet assemblers/disassemblers

Packet assemblers/disassemblers (PADs) were developed to enable non-packet-forming terminal equipment to communicate over packet networks. Originally, PADs were designed to convert the asynchronous start–stop data flow from two or more terminal devices into an X.25 data stream, in effect, performing multiplexing and protocol conversion.

Since the introduction of PADs during the 1970s, their functionality as well as physical appearance has considerably changed. In the 1970s, PADs were based upon the use of minicomputer technology and resembled in physical size a small desk. By the 1990s, PADs were obtainable as adapter cards that could be inserted into the system unit of a personal computer. Along with a change in the physical characteristics of PADs was a change in their capability. Since their original introduction as asynchronous-to-X.25 conversion devices, many vendors added support for other protocols, including IBM bisynchronous 2780, 3780 and 3270, as well as HDLC and its vendor-specific derivatives.

Figure 1.10 illustrates the operation, utilization, and typical placement of PADs. At the top of that illustration, a PAD is located at a customer's premises and serves both as a line-sharing device and a protocol conversion device. In the lower portion of Figure 1.10, the PAD is located at a packet network node.

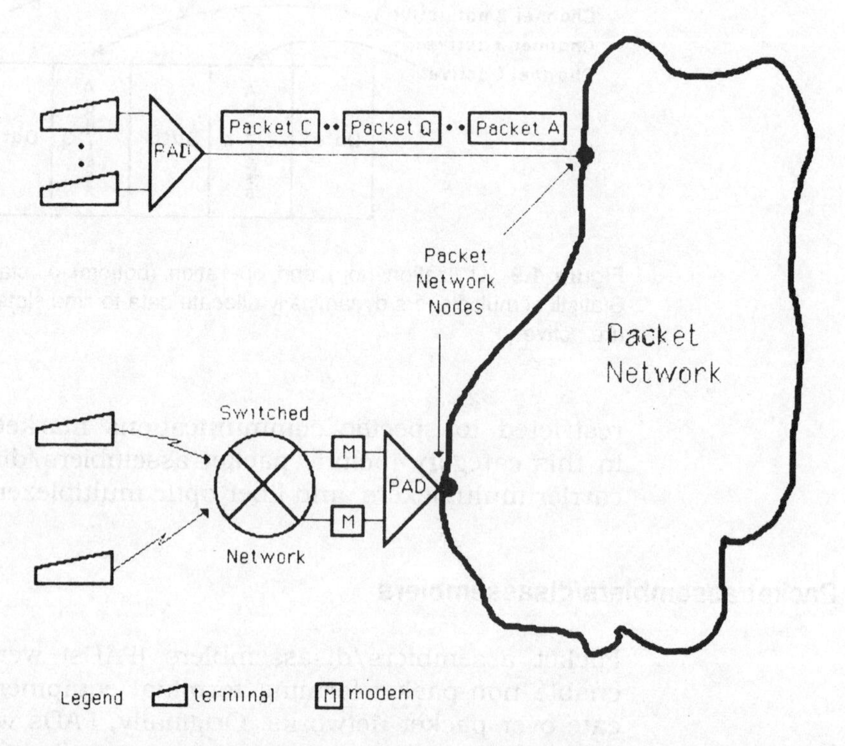

Figure 1.10 PAD placement. Packet assemblers/disassemblers (PADs) can be located at a customer's premises or at a packet network node.

Normally, a PAD is located at a subscriber's premises when the subscriber has two or more terminal devices, each requiring a fairly large amount of connect time. When a subscriber has only one terminal device, or a few with nominal transmission requirements, it is typically more economical for users to dial a public PAD available to all packet network subscribers. For both methods of PAD placement, non-packet terminal devices become compatible with the packet network through the functionality of the PAD. Today, PADs are one of the most popular types of multiplexers since their use parallels the growth in the use of packet networks.

T-carrier multiplexers

Until the mid-1980s, the multiplexing of North American T1 and European E1 circuits was essentially restricted to telephone companies routing multiple digitized voice conversations on one T1 or E1 circuit routed between telephone company central offices. This restriction was due to the unavailability of commercial T1 and E1 facilities and lasted until AT&T in the United States and British Telecom in the United Kingdom tariffed those facilities for commercial usage. When this occurred, the cost structure was such that the usage of only a small portion of the 1.544 Mbps data rate of the T1 carrier in the United States or the 2.048 Mbps rate of the equivalent E1 carrier in Europe normally resulted in significant economic savings in comparison to the use of multiple analog voice and data lines.

The economics associated with using a T-carrier resulted in a demand for multiplexing equipment that could multiplex both voice and data. Most T-carrier multiplexers originally introduced were marketed with an optional pulse code modulation (PCM) module, whose use resulted in the digitization of a voice conversation at a 64 kbps data rate. Due to advances in voice digitization techniques, more modern T-carrier multiplexers normally provide the end-user with an assortment of optional voice digitization modules. Typical data rates resulting from the use of these modules are 16 kbps, 32 kbps, 48 kbps, and 64 kbps, with the reconstructed voice quality slightly increasing as the digitization rate increases. Since a 32 kbps digitization rate doubles the voice-carrying capacity of a T-carrier multiplexer with a slight to unperceivable loss in voice quality, many end-users have purchased digitization modules for their multiplexers that convert voice to that digital data rate.

Today, the T-carrier multiplexer represents the major platform by which organizations can construct integrated voice and data networks. From a few commercial sales during its introductory year, the acceptance of this product can be judged by its current market. Today, the sale of T1/E1 multiplexers exceeds $500 million per year and represents more than half the value of all types of multiplexers sold.

Fiber optic multiplexers

The development of the fiber optic multiplexer resulted from the merging of time division multiplexing technology with fiber optic transmission technology. In the late 1970s, several vendors noted a demand for the transmission of data in the form of light energy to alleviate the problems and expense associated with the transmission of data in the form of electricity.

One major problem facing end-users was the transmission of information in hazardous areas, such as petroleum refineries, where a spark could result in disaster. A second major problem area was the routing of data within a building, where building codes required the construction of expensive conduits to shield conventional data cables. Since fiber optic transmission resulted in information flowing in the form of light energy, it was suitable for use in hazardous areas and did not require the use of a conduit. In fact, if a conduit was previously installed, the fiber optic cable could be taped to the outside of the conduit, enabling it to follow the route of the conduit without affecting the capacity of the metal housing to contain additional electrical cables.

Due to the advantages of using fiber optic cable, many organizations began to use fiber optic transmission within their data centers during the late 1970s and early 1980s. At the same time, communications carriers began to use fiber optic cable to carry tens of thousands of simultaneous voice conversations between major metropolitan areas. The key to the successful use of fiber optic cable was the development of two different types of fiber optic multiplexers—one is primarily used by both communications carriers and commercial organizations and is discussed in detail in this book, while the second is primarily in a research stage and is briefly described in this book.

The fiber optic multiplexer used primarily by communications carriers and commercial organizations is based upon the use of conventional TDM technology. Here vendors merged the use of a light emitting diode (LED) or low-power laser with a photodetector for the transmission and reception of light energy with the electronics of the TDM. Doing so enables many data sources to share the high data transmission capacity of the fiber optic cable. Although this type of fiber optic multiplexer is also used by communications carriers, at the time this book was written they were beginning to use, as well as researching, the use of multiplexers that operated based upon the separation of a light source into channels based upon the frequency of the light.

Regardless of the type of fiber optic multiplexer used, its use has numerous advantages. First and probably foremost is the ability of the multiplexer to enable many data sources to share the use of a fiber optic cable. This in turn enables relatively high data transmission rates to be achieved, with each multiplexed data source becoming immune to electromagnetic interference as it travels down the cable. In addition, since energy is carried in the form of light, the cable can be routed outside of conduits—an important consideration when cabling space within a building is at a premium.

1.3 FUTURE DEVELOPMENTS

At the time this book was prepared, there were a number of new technologies whose development and implementation could be expected to have a considerable effect upon multiplexer technology. New transmission technologies include frame relay, cell relay, and Synchronous Optical Network (SONET). The first two transmission technologies have begun to affect the development of T-carrier multiplexers as manufacturers have modified their equipment to obtain compatibility with newly evolving standards. The third transmission technology affects the development of fiber optic multiplexers being developed to support the transmission requirements of SONET. In addition to the previously mentioned transmission technologies, several new features were on the verge of being incorporated into multiplexers. Recognizing that bridging and routing between local area networks (LANs) are in many instances designed to interconnect widely dispersed LANs, vendors were investigating adding these features to statistical and T-carrier multiplexers. In relevant chapters in this book we will examine these transmission technologies and their current and projected impact upon multiplexer technology.

2

FREQUENCY DIVISION MULTIPLEXERS

In this chapter we will focus our attention upon the operation and utilization of frequency division multiplexers and the frequency division multiplexing process. Although frequency division multiplexing is an analog-based technology whose use in communications networks is rapidly being replaced by more modern digital-based systems, it will probably remain in limited use throughout the remainder of this century. In addition, there are several functions associated with the use of FDM that could prolong its utilization into the next century. Thus, our examination of FDM will include several examples that illustrate the features of this multiplexing technology that could extend its useful life.

Readers should note that our discussion of FDM in this chapter does not include wavelength division multiplexing (WDM). The latter references a technique similar to FDM used in optical fiber systems in which the optical transmission spectrum is subdivided into discrete subchannels based upon frequency. Although WDM is based upon FDM, it is implemented using discrete digital components and is more appropriately discussed in a later chapter which covers fiber optic multiplexers.

2.1 TYPES OF FREQUENCY DIVISION MULTIPLEXING

Although we may not realize it, we probably encounter FDM every day of our lives. Whenever we turn on a television or radio we are working with FDM. Here each television channel or radio station is assigned to a discrete frequency that is separated by frequency from other television channels or radio stations. In this particular example the atmosphere is the transmission medium. However, those of us with cable TV access similarly experience FDM; in this situation the coaxial cable is the transmission medium. In radio transmission each station modulates a carrier of frequency assigned to the station. Here amplitude modulation is used for AM stations, while frequency modulation is used for FM stations. Your radio tuning circuits enable you to separate one AM or FM signal from all other signals as you tune the radio to a specific frequency.

Since the primary focus of this book is upon the use of multiplexers to network voice and data transmissions, we will exclude television and radio systems from our examination of different types of frequency division multiplexing. Doing so will enable us to focus our attention upon two specific types of multiplexers—those used by communications carriers to place multiple-voice conversations on wideband transmission facilities and those used by commercial organizations as a data multiplexer.

2.2 COMMUNICATIONS CARRIER SYSTEMS

Modern analog FDM techniques used by communications carriers were developed during the 1930s and have been standardized by the Consultative Committee for International Telephone and Telegraph (CCITT) as well as by other standard organizations. At that time, engineers developed equipment which enabled voice conversations to be filtered into a discrete range of frequencies as well as to shift the frequency range upward and downward by frequency.

When a communications carrier uses FDM for the multiplexing of voice conversations onto a common circuit, each conversation is filtered at a lower level of approximately 300 Hz and an upper level of 4 kHz. Next, the resulting passband of each conversation is shifted upward in frequency by a fixed amount of frequency. This frequency shifting places the voice conversation into a predefined channel of the FDM multiplexed channel. At the opposite end of the circuit, another FDM demultiplexes each voice conversation by shifting the frequency spectrum of each conversation downward in frequency by the same amount of frequency as it was previously shifted upward.

As previously mentioned, the primary use of FDM equipment by communications carriers was to facilitate a large number of simultaneous voice conversations on a common circuit routed between two or more carrier offices. The actual process for allocation of the bands of frequencies to each voice conversation was standardized by the CCITT based upon a grouping of 12 voice channels at 4 kHz to provide a combined analog signal of 48 kHz. Known as a standard group, this group can be further multiplexed into a standard supergroup representing five standard groups or a standard mastergroup representing five standard supergroups. Thus, CCITT FDM recommendations govern the assignment of 12, 60 and 300 voice channels on analog wideband circuits.

CCITT FDM wideband recommendations

In this section, we will examine CCITT FDM wideband recommendations that govern the multiplexing of voice conversations onto analog wideband facilities. These recommendations should not be confused with other CCITT FDM recommendations described later in this chapter which govern the subdivision of a single analog voice line into subchannels, enabling multiple data transmissions to be simultaneously carried on one voice circuit.

The standard group

The standard group as defined by CCITT recommendation G.232 is based upon a grouping of 12 voice channels at 4 kHz to provide a combined analog line signal of 48 kHz. Here the filtering of voice signals below 300 Hz for each conversation results in the generation of a 300 Hz guardband which separates each voice conversation. The resulting group that represents 12 voice conversations is placed into the 60 to 108 kHz frequency region as illustrated at the top of Figure 2.1. The standard group can be considered as the first level of frequency division multiplexing under CCITT recommendation G.232.

Figure 2.2 illustrates the method by which the CCITT standard group is generated. Here each voice channel passes through a low-pass filter to remove all frequency components under 300 Hz. The carrier frequency generator produces discrete carrier signals from 64 kHz to 108 kHz in 4 kHz increments. Each carrier signal is used to modulate the resulting low-pass filtered voice conversation, and the resulting

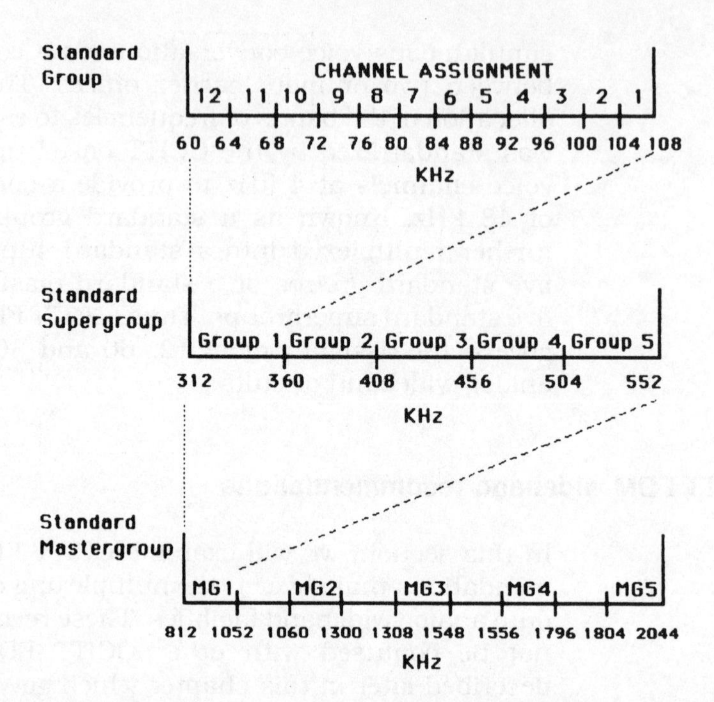

Figure 2.1 Standard CCITT FDM groups. CCITT FDM recommendations govern the assignment of 12, 60, and 300 voice channels on wideband analog circuits.

modulated signal is then again filtered to reside within a specific frequency band. This second filtering process removes the components of the modulated signal and ensures that one multiplexed signal does not interfere with another multiplexed signal. The pilot frequency can be considered as a reference frequency which is used for both modulation and demodulation. Although Figure 2.2 illustrates one-way transmission for simplicity, in actuality a second group of equipment is used which first filters received signals into their appropriate frequency range, demodulates the filtered signal using the previously described carrier frequencies produced by the carrier frequency generator, and then uses a low-pass filter to remove any frequencies below 300 Hz. Thus, two-way conversations are then supported.

The standard supergroup

By employing a second modulation step with a different carrier for each group and using five carriers, a standard supergroup is formed. The standard supergroup defined by

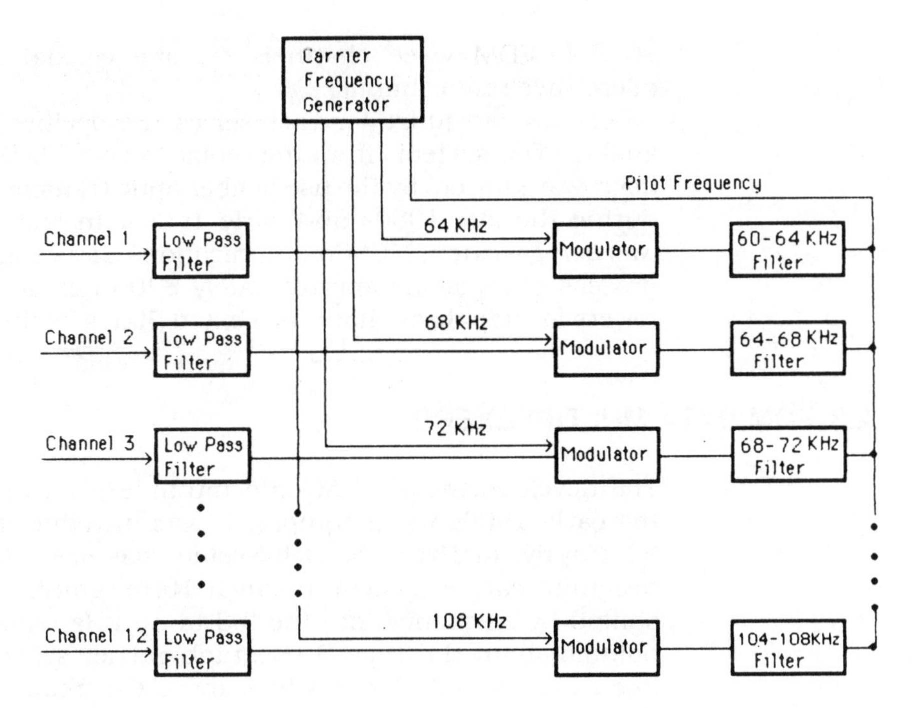

Figure 2.2 CCITT standard group formation. A standard group contains 12 channels placed into the 60 to 108 KHz frequency range.

CCITT recommendation G.241 contains five standard groups which represent 60 voice channels. The standard supergroup occupies 240 kHz of frequency between 312 and 552 kHz as illustrated in the middle portion of Figure 2.1.

The standard mastergroup

The third CCITT FDM recommendation, known as the standard mastergroup, can be considered as the top of the FDM hierarchy. The standard mastergroup is formed through the use of five carriers, each of which shifts a standard supergroup in frequency. Since each supergroup consists of 60 voice channels, the mastergroup contains a total of 300 voice channels. The standard mastergroup occupies the frequency band from 812 to 2044 kHz, as illustrated in the lower portion of Figure 2.1.

In addition to the three FDM multiplexing levels (four when referenced to a single voice channel), other systems were developed that added as many as three additional modulation levels. Such systems were developed to provide as many as

10 800 FDM voice channels on one coaxial cable or one microwave radio channel.

Readers can judge for themselves the decline in the use of analog FDM systems from the replacement of large numbers of microwave towers by the use of fiber optic transmission systems during the late 1980s and early 1990s. In fact, as this book was being written, MCI Communications Corporation was in the process of replacing approximately $200 million of microwave towers located throughout the United States by fiber optic cable.

2.3 FDM DATA MULTIPLEXERS

The development of FDM data multiplexers can be traced to the early 1900s when equipment was introduced to multiplex telegraphy. In 1915, the Bell System was operating its Type B telegraph carrier system in which 10 telegraph channels were shifted by frequency into the 3.3 to 10 kHz region. What was notable about the Type B telegraph carrier system is that the lower limit of 3.3 kHz was just above the frequency of a voice conversation, enabling what is referred to as 'data over voice' to occur in which one or more low-speed data transmission channels could be simultaneously carried on the same circuit with a voice conversation through the use of FDM.

Today, data over voice systems are in use on many international circuits as well as in intrabuilding communications systems. Concerning the former, many organizations with international offices interconnected by leased lines use data over voice as a mechanism to obtain a 'free' data transmission capability on an expensive voice circuit. Concerning the latter, many modern buildings are wired with coaxial cable or fiber optic cable which enables voice, data, and even video to be carried on a common cable through the use of FDM. Since those cables have a much wider bandwidth than twisted pair, higher data rates are supported.

Carrier telegraph

Early carrier telegraph systems subdivided a voice channel into several independent channels by frequency. The FDM system simply turned the carrier tone on and off to represent binary ones and zeros. Since the carrier's signal power was simply changed between a maximum amplitude and zero, this technique was referred to as AM FDM or AM carrier telegraph. Such carrier telegraph systems were standardized under the CCITT R.31 recommendation as well as implemented by several manufacturers as proprietary systems. Between 10 and 24 telegraph channels were obtained by subdividing a 3 kHz voice channel, with the transmission speed per channel usually limited to 50 to 135 bps.

Under the CCITT R.31 recommendation, the voice channel is subdivided into 24 telegraph channels, with each subchannel operating at 50 bps. Table 2.1 lists the lower frequency in each of the 24 bands established under the CCITT R.31 recommendation.

Table 2.1 CCITT recommendation R.31 frequency assignments

Channel	Lower frequency of band (Hz)	Channel	Lower frequency of band (Hz)
1	420	13	1860
2	540	14	1980
3	660	15	2100
4	780	16	2220
5	900	17	2340
6	1020	18	2460
7	1140	19	2580
8	1260	20	2700
9	1380	21	2820
10	1500	22	2940
11	1620	23	3060
12	1740	24	3180

Early carrier telegraph systems using the AM technique were very susceptible to transmission errors. This was because noise could provide a false amplitude level which the AM FDM receiver could not distinguish from an on or marking carrier tone. This problem resulted in the development of carrier telegraph multiplexing systems in which the carrier tone's frequency was varied in conjunction with the data in place of varying the amplitude. Such systems transmit two tones, one to represent a binary one or mark, while the second tone represents a binary zero or space. In between each bit the carrier continues to stay at the prior represented frequency, enabling the carrier tone to remain at its full power level prior to shifting frequencies. This technique is also known as frequency shift keying (FSK) and was incorporated into low-speed modems.

The use of FSK in carrier telegraph FDM systems considerably reduced the problem of noise being interpreted as data in AM FDM systems and formed the basis for the development of modern FDM data multiplexers. When used with carrier telegraph the use of frequency modulation was referred to as FM voice frequency carrier telegraph or FMVFCT.

Modern FDM data multiplexers

As previously discussed, the advantage of frequency modulation over amplitude modulation resulted in the preferred use of FM FDMs for carrier telegraph systems. This method of FDM was incorporated into modern FDM data multiplexers that were originally developed to support the computer communications requirements of commercial organizations and government agencies during the late 1950s and early 1960s. At that time, a mechanism was required to allow terminal users to share the use of leased lines routed from remote locations to a centralized mainframe computer. During that period, the preferred terminal was the Teletype Corporation Model 33, a line-by-line asynchronous teleprinter which was originally designed for use on AT&T's Teletypewriter Exchange Service (TWX). This terminal was readily available for use with computers and formed the basis for most computer communications networks established in North America during the late 1960s and the early 1970s. In fact, the popularity of this terminal resulted in hundreds of manufacturers producing TTY-compatible products, a term coined to refer to an asynchronous terminal that used the ASCII code and transmitted and received data on a line-by-line basis.

Through the use of frequency modulation, the modern FDM data multiplexer uses pairs of frequencies within each channel to represent the ones and zeros of each data source. The splitting of the line into smaller segments called data bands or derived channels and the use of a guard band between channels to prevent signal interference between channels was previously illustrated in Figure 1.6.

Physically, an FM FDM contains a channel set for each data channel as well as common logic which governs the operation of the multiplexer. Each channel set contains a transmitter and receiver tuned to specific pairs of frequencies. The transmitter shifts tones by frequencies to represent marks and spaces, while the receiver converts received tones at specific frequencies into marks and spaces. Figure 2.3 illustrates in block diagram format a central site FM FDM as well as several terminals remotely located from the central site that share access to the central site via a common frequency division multiplexed line.

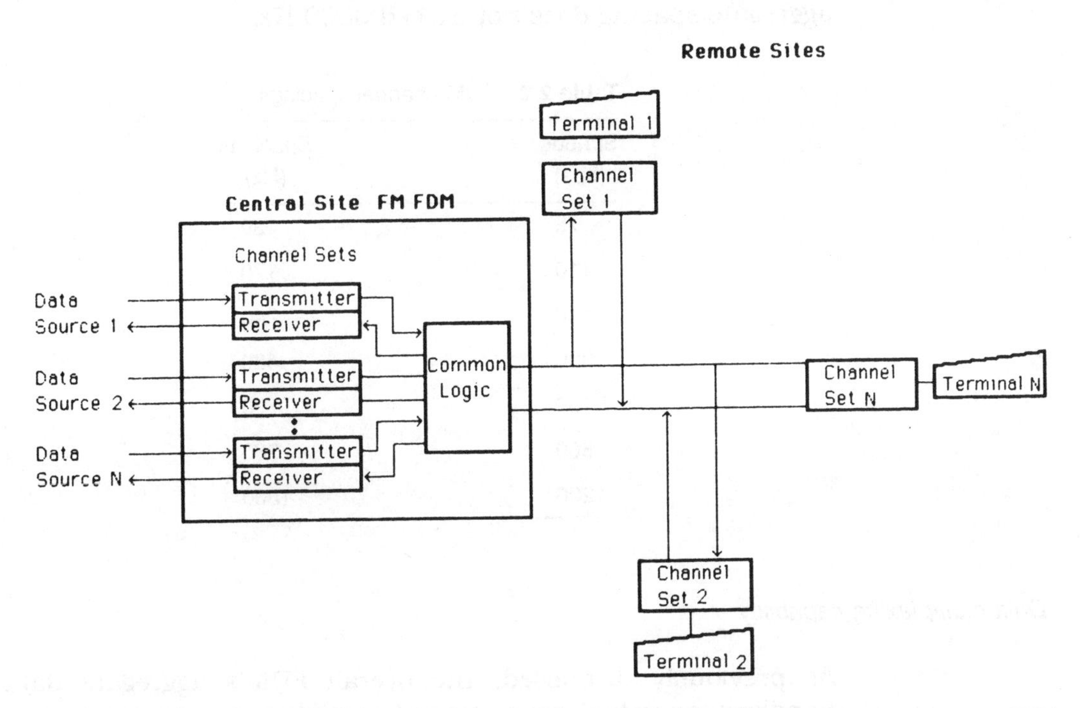

Figure 2.3 FM FDM operation. The channel sets used in FM FDM can be considered as FSK modems, hence no modems are required.

In examining Figure 2.3, note that each channel set in effect functions as an FSK modem. Thus, channel sets can be considered to be FSK modems. Also note that since the bandwidth of the circuit linking the central site to the remote locations is subdivided into discrete frequency bands, FDM enables terminals at the same or different locations to share the use of the circuit without poll and select software. Thus, FDM provides users with an inherent multipoint networking capability without requiring specialized software.

In FDM, the width of each frequency band determines the transmission rate capacity of the channel, with the total bandwidth of the line the limiting factor in determining the total number or mixture of channels that can be serviced.

Although a multipoint operation is illustrated in Figure 2.3, FDM equipment can also be utilized for the multiplexing of data between two locations on a point-to-point circuit. Typical FDM channel spacings required at different data rates are listed in Table 2.2. Since a voice-grade line is limited to approximately 300 Hz bandwidth, you are limited to multiplexing six 300 bps, three 600 bps, or any other mixture of data sources whose aggregate spacing does not exceed 3000 Hz.

Table 2.2 FDM channel spacings

Speeds (bps)	Spacings (Hz)
75	120
110	170
150	240
300	480
450	720
600	960
1200	1800

Data multiplexing capability

As previously discussed, the overall FDM's aggregate data handling capacity depends upon the mixture and operating rate of derived subchannels. Figure 2.4 illustrates an FDM allocation chart that can be used to compute the mixture of data channels that can be transmitted via a single voice-grade circuit when FM frequency division multiplexing is employed. The referenced chart is based upon data channel spacing standards formulated by the CCITT. Although the chart indicates the maximum speed in baud, under FM each bit is encoded as a single signal change. Thus, the term 'baud' on the chart is synonymous with a data rate expressed in bps.

With the development of terminals operating at speeds that were not multiples of 75 baud, such as 134.5 bps teleprinters,

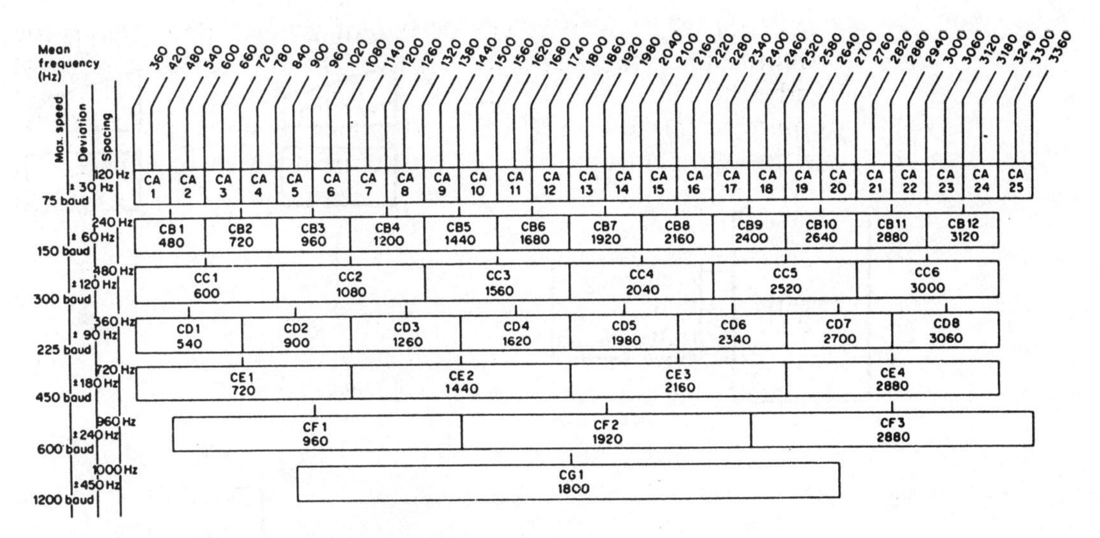

Figure 2.4 CCITT FDM subchannel allocations.

a number of vendors developed FDM equipment tailored to make more efficient use of voice-grade circuits than permitted by CCITT standards. Other vendors simply used a larger channel spacing which, although inefficient, enabled non-standard data rates to be multiplexed. In fact, this is another advantage obtained from the use of FDM—once a channel spacing is assigned for one data rate, other data rates under that rate can use that channel. Another advantage obtained from the use of FDM is its inherent code transparency. Once a data band is set, any terminal operating at that speed or less can be used on that channel without concern for the code of the terminal. Thus, a channel configured to carry 300 bps transmission could also be used to service an IBM 2741 terminal transmitting at 134.5 bps or a Teletype 110 bps terminal.

Since the physical bandwidth of the line limits the number of devices which may be multiplexed, FDM is primarily used for multiplexing low-speed asynchronous terminals. Since the channel sets modulate the line at specific frequencies they in effect function as FSK modems. Normally, an FDM consisting of an enclosure containing several channel sets is installed at a computer site, with each channel set connected to a low-speed computer port. Thus, the left portion of Figure 2.5 would correspond to the installation of FDM equipment at a computer site.

At each remote location, a channel set provides the necessary interface between the terminal at that location and the leased

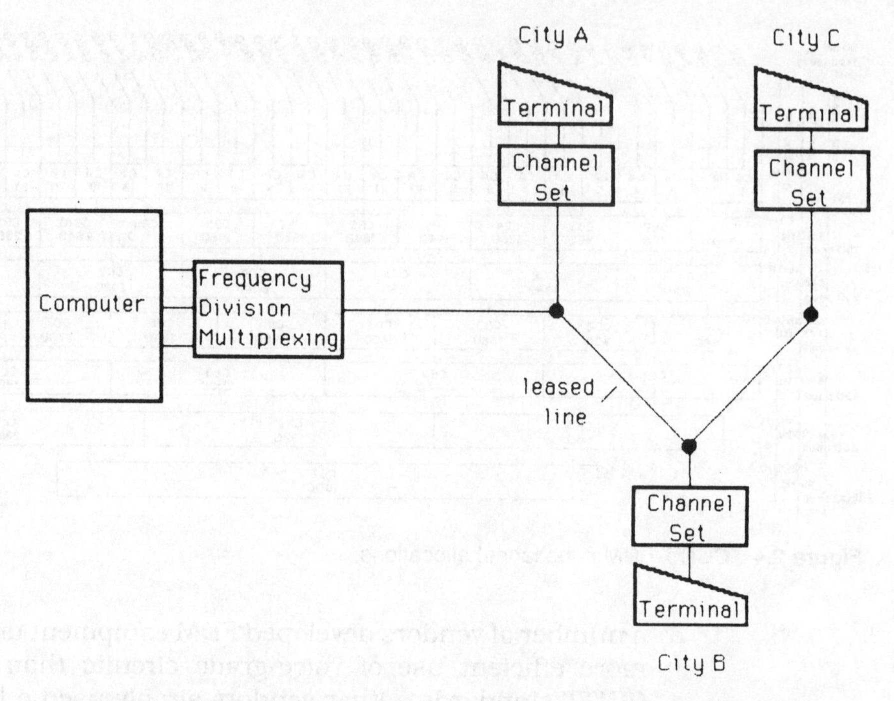

Figure 2.5 FDM multipoint support. Frequency division multiplexing permits multipoint circuit operations. Each terminal on an FDM multipoint circuit is interfaced through the multiplexer to an individual computer port.

line. This is illustrated in the right portion of Figure 2.5, with locations labeled City A, City B, and City C containing channel sets.

When using FDM, individual data channels can be picked up or dropped off at any point on a telephone circuit. This characteristic permits FDM to provide a multipoint or multidrop capability without requiring poll and select software. This in turn can provide organizations with the ability to considerably reduce line charges since a single circuit can be used to support multiple terminals at different locations without specialized software. Each remote terminal to be serviced needs only to be connected to an FDM channel set which contains filters that separate the line signal into the individual frequencies designated for that terminal. Guard bands of unused frequencies are used between each channel frequency to permit the filters a degree of tolerance in separating out the individual signals.

FDM normally operates in a full-duplex transmission mode on a four-wire circuit by having all transmit tones sent on one pair of wires and all receive tones returned on a second pair. However, FDM can also operate in the full-duplex mode on a two-wire line; this can be accomplished by having the transmitter and receiver of each channel set tuned to different frequencies. For example, with 16 channels available, one channel set could be tuned to channel 1 to transmit and channel 9 to receive, while another channel set would be tuned to channel 2 to transmit and channel 10 to receive. With this technique, the number of data channels is halved. However, the cost difference between a four-wire and a two-wire circuit may justify its use if your organization has only a small number of terminals to service.

FDM utilization

As previously discussed, one key advantage obtained from the use of FDM is the ability it provides in obtaining a multipoint capability without requiring poll and select software. This capability can minimize line costs since a common circuit, optimized in routing, can now be used to service multiple terminal locations. An example of FDM equipment used on a multipoint circuit is shown in Figure 2.5, where a three-channel FDM is used to multiplex traffic to and from terminals located in three different cities. In this example, the channel set at each location contains a transmitter and receiver tuned to specific pairs of frequencies. Thus, frequency modulated data destined for each specific location will not be heard by the channel set receivers at other locations. Similarly, the transmitter at each location will shift between a pair of predefined frequencies that the multiplexer at the computer site recognizes as being assigned to a specific channel set in the FDM. Hence, data is demodulated and routed to the computer port associated with a specific terminal location.

In contrast to poll and select multipoint line operations, where one computer port is used to transmit and receive data from many buffered terminals with unique addresses connected to a common line, the use of FDM for multipoint operations as shown in Figure 2.5 requires one computer port for each terminal. However, such terminals do not require a buffer area and addressability, nor is poll and select software required to operate in the computer. When buffered, addressable terminals and poll and select software are available, FDM can be used to considerably increase the number of terminals and terminal

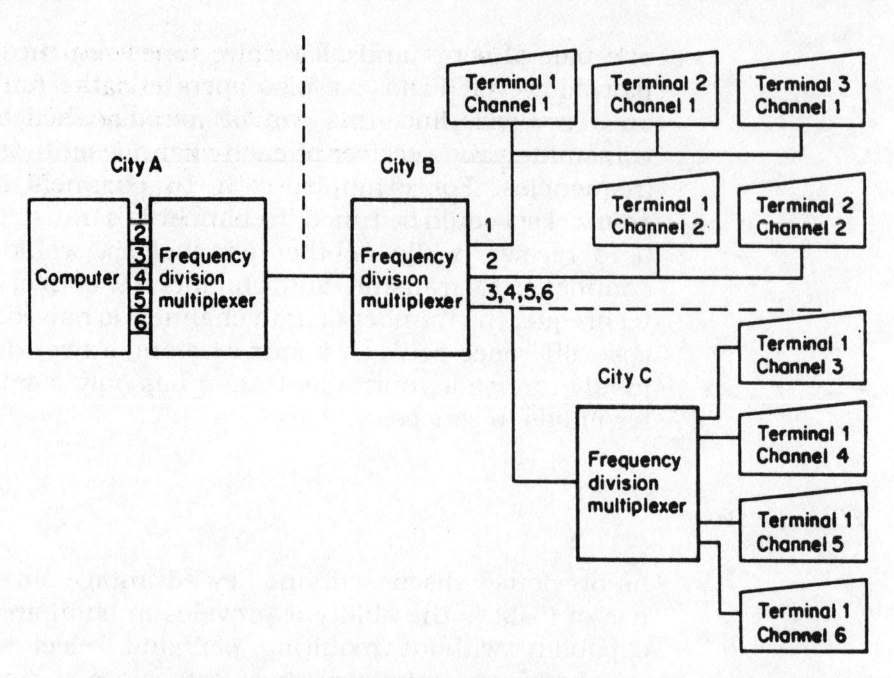

Figure 2.6 FDM can intermix polled and dedicated terminals in a network. Of the six channels used in this network, channels 1 and 2 service a number of polled terminals, while channels 3 through 6 are dedicated to service individual terminals.

drops that can be serviced over a common multidrop line. Figure 2.6 illustrates an example of how polling by channel can be used with FDM systems. In this example, channels 1 and 2 are each connected to a number of relatively low-traffic terminals which are polled through the multiplexer system. Terminals 3 through 6 are presumed to be higher-traffic stations and are thus connected to individual channels of the FDM via the use of individual channel sets.

Today the primary use of FDMs for data transmission is in support of a large number of geographically dispersed low-speed terminals, where the volume of transmission is frequent, but minimal. For example, trucking companies that have hundreds of depots interconnect low-speed terminals by multidrop circuits whose operation combines poll and select software with channel allocation to support 30 to 50 terminals on one circuit. Since such organizations use their terminals to denote the arrival and departure of trucks that may next be on the highway for hours, speed of transmission is not a primary concern. A second area where FDM can be expected to continue to be used is in providing data over voice on international circuits, in which one or a few low-speed data circuits are added to a voice circuit through the use of frequencies beyond 3000 Hz. Since this utilization is expected to provide a limited market for FDM for the foreseeable future, to paraphrase Mark Twain and Yogi Berra, the demise of FDM ain't here yet!

3

TIME DIVISION MULTIPLEXING

In this chapter, we will focus our attention upon the process of time division multiplexing and the operation and utilization of time division multiplexers (TDMs). In doing so we will first compare frequency division multiplexing to time division multiplexing to provide readers with a frame of reference between the two technologies. This will be followed by an examination of TDM operations, TDM applications, and the functionality and use of other types of communications devices that incorporate TDM technology.

3.1 FDM VERSUS TDM

In the FDM technique, the bandwidth of the communications line serves as the frame of reference for multiplexing. Here the total bandwidth is divided into discrete channels consisting of smaller segments of the available bandwidth, each of which is used to form an independent transmission facility. In the TDM technique, the aggregate transmission capacity of the line is the frame of reference, since the multiplexer provides a method of transmitting information from many data sources over a common line by interleaving each data source in time.

A basic TDM divides the aggregate transmission capacity governed by the speed of the transmission facility connected to the multiplexer into time slots. Each data source interfaced to the TDM is given one or more time slot intervals for its exclusive use, with the number of time slots assigned to a specific data source based upon the operating rate of the data source. Thus, at any point in time the signal from only one data source flows on a time division multiplexed line. In comparison, when frequency division multiplexing is used, many data sources can be simultaneously carried on a circuit, since the circuit is subdivided into independent channels by frequency.

Although time division multiplexing is limited to placing information from one data source on a line at any point in time, its overall operation is governed by the operating rate of the multiplexed circuit. When TDM systems became available for commercial use in networks during the 1960s, they immediately provided a data multiplexing capacity beyond that available from the use of FDM. This was due to the availability of 4800 bps leased line modems which governed the mixture of lower-speed data sources that could be multiplexed. In comparison, as noted in Chapter 2, FDM is limited due to the proportional channel spacing in Hz required by different data rates. Thus, a maximum of six 300 bps data sources could be multiplexed using FDM, while approximately eighteen 300 bps data sources could be multiplexed using a TDM connected to a 4800 bps modem. Due to the additional data multiplexing capacity of TDM, such systems rapidly replaced FDM in most computer networks and became the preferred method for data multiplexing.

The basic method of time division multiplexing described in this chapter has been incorporated into stand-alone multiplexers, several types of high-speed synchronous modems and data service units. In addition, this method of multiplexing formed the basis for the development of statistical multiplexers, packet assemblers/disassemblers, T-carrier multiplexers, and fiber optic multiplexers, each of which are discussed in the following chapters in this book.

3.2 TDM OPERATION

The fundamental operating characteristics of a time division multiplexer are illustrated in Figure 3.1. Here each data source is connected to the multiplexer through an input/output (I/O) channel adapter.

I/O channel adapter

The I/O channel adapter includes a small buffer as well as control logic and provides a speed interface between the operating rate of an attached data source and the electronic

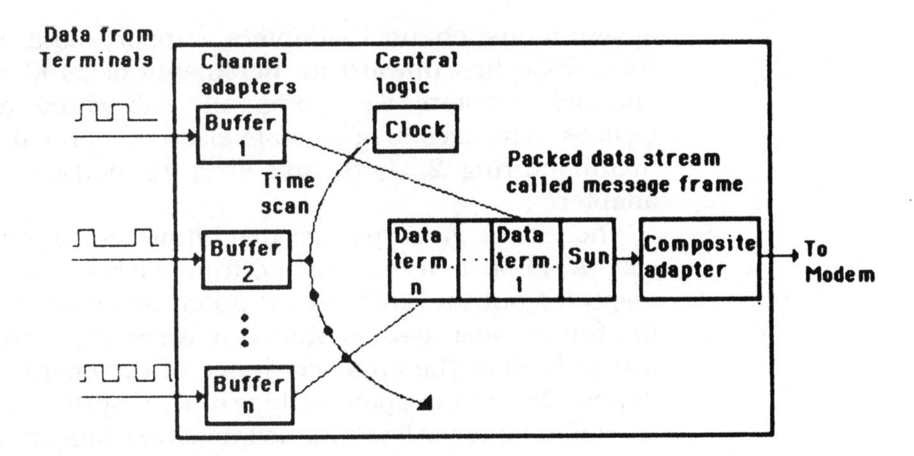

Figure 3.1 Time division multiplexing. In TDM, data is first entered into each channel adapter buffer area at a transfer rate equal to the device to which the adapter is connected. Next, data from the various buffers is transferred to the multiplexer's central logic at the higher rate of the device for packing into a message frame for transmission.

operating rate of the multiplexer. In actuality, the buffers in each channel adapter usually consist of a pair of double buffers, with one set of buffers used for data flowing into the multiplexer, while the second pair of buffers is used for transmitting demultiplexed data to an attached device. The reason for a pair of buffers for input and output is that one buffer operates at the data rate of the attached device, whereas the second buffer operates at the electronic speed of the multiplexer. For example, if the buffer is 8 bits in size, each character flows into the buffer at the operating rate of the attached device. Once filled, the contents of the buffer are shifted to a second buffer, making the first buffer available for the next 8-bit character. The central logic of the multiplexer includes clocking circuitry which governs the scanning and emptying of the second buffer in the channel adapter, whose contents are packed into a data stream known as a message frame.

Data is shifted from the second buffer in each channel adapter in a round-robin fashion into the message frame, with a synchronization character (Syn) prefixed to each frame to provide synchronization between multiplexers. The operating rates supported by a channel adapter are dependent upon the type of adapter used. Most asynchronous channel adapters support data rates from 110 to 9600 or 19 200 bps, whereas synchronous channel adapters support data rates ranging from 2400 bps upward in increments of 2400 bps. Although channel adapters were originally fabricated as single-port devices, advances in solid state electronics resulted in vendors manufacturing 2, 4, 8, and even 16 ports on one channel adapter.

When data is output or demultiplexed, a reverse process occurs. First data is extracted from each frame and routed to a specific output buffer based upon the position of the data in the frame. Next, data is shifted to a second output buffer and transmitted to the attached device at the operating rate of the device. Depending upon the type of TDM system, the buffer area in each adapter will accommodate either bits or characters.

Central logic

The central logic of the TDM contains controlling, monitoring, and timing circuitry which facilitates the passage of individual data sources to and from the high-speed transmission medium. The central logic also generates a synchronizing pattern, which is used by a scanner circuit to interrogate the buffer area in each channel adapter in a predetermined sequence, normally following a round-robin sequence. The data in each buffer, either a bit or a character, is then packed into a predefined area of memory in the TDM until all data sources are operated upon during a scanning cycle. Once the scanning cycle is completed, the resulting packed data stream is shifted into the multiplexer's composite adapter, from which it is transferred onto the connected circuit.

Composite adapter

The composite adapter contains a dual buffer pair for transmission and reception of data, with each buffer pair functioning similar to I/O channel adapter buffers. That is, data in the form of a message frame is transferred into one buffer at the electronic speed of the multiplexer and then rapidly transferred to a second buffer from which it is emptied at the speed associated with the transmission line. In actuality, the synchronous data stream is either transmitted at the clocking rate of an attached modem or data service unit (DSU) if the composite adapter is configured to external timing. If the composite adapter supports and is set to internal timing, then data is transmitted based upon the setting of the internal clocking speed of the adapter.

The message frame

The packed data stream, which is emptied from the composite adapter, is also referred to as the multiplexer's message frame. If there is no activity generated by a data source when its associated channel adapter is scanned, the TDM circuitry fills the position assigned to the data source in the message frame with a pad character. This pad character is normally the NUL character in a character set and is required to maintain the position of data sources in the message frame, since demultiplexing occurs at the opposite end of the circuit connecting two multiplexers based upon the position of data within the message frame.

Figure 3.2 illustrates a simple example of the multiplexing and demultiplexing process for a two-part TDM. In this example the absence of activity on port 1 during a scan is indicated by a dashed (—) line, which is converted into a pad character indicated by a triangle (△) by the multiplexer for insertion into the multiplexing frame. The receiving multiplexer at location 2 then converts the pad character into a period of inactivity indicated by the dashed line in the demultiplexed output. Hence, the pad character insertion and removal enables demultiplexing to occur by the position of data in the multiplexing frame.

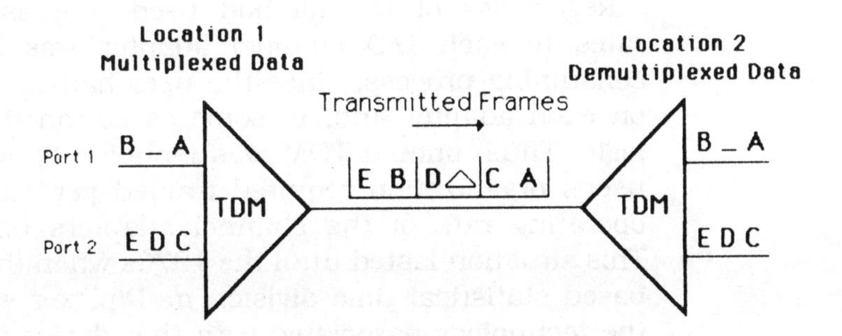

Figure 3.2 TDM multiplexing and demultiplexing. Inactivity (—) on a port is denoted by the insertion of a pad character (△) into the multiplexing frame.

The multiplexing interval

When operating, a TDM transmits and receives a continuous data stream known as a message train, regardless of the activity of the data sources connected to the multiplexer. The message train is formed from a continuous series of message frames which represents the packing of a series of input data streams. Each message frame is prefixed with one or more synchronization characters, followed by a number of basic multiplexing intervals whose number is dependent upon the model and manufacturer of the device.

The basic multiplexing interval can be viewed as the first level of time subdivision. The interval is established by determining the number of equal sections per second required by a particular data source. Thus, the multiplexing interval is the time duration of one section of the message frame.

When TDMs were first introduced, the section rate was established at 30 sections per second, which produced a basic multiplexing interval of 0.033 s or 33 ms. Setting the multiplexer interval to 33 ms made the multiplexer directly compatible to a 300 bps asynchronous data source which transmits data at up to 30 characters per second. With this interval, the multiplexer was compatible with 150 bps and 110 bps data sources, since the basic multiplexing interval was a multiple of those asynchronous data rates. For servicing higher-speed devices, most multiplexers assign multiple intervals to higher operating rate data sources, with, as an example, four intervals assigned to a 1200 bps data source. Other multiplexers are designed with a higher section rate, with 120 sections per second used to service asynchronous data streams up to 1200 bps. Some multiplexers use both techniques, raising the section rate and assigning multiple intervals to higher-speed data sources.

Regardless of the method used, the assignment of data rates to each I/O channel adapter was originally a time-consuming process, since the user had to set program plugs on each adapter and, in some cases, on the device's central logic. Thus, once a TDM was installed by a vendor, the end-user's organization required trained personnel to change the operating rate of the channel adapters on the multiplexer. This situation lasted until the 1970s when the microprocessor-based statistical time division multiplexer was developed and the technology associated with that device, including either a control port that could be cabled to an asynchronous terminal or a front panel display which enabled users to change channel speeds, was incorporated into TDM equipment.

TDM techniques

Two techniques are used in TDM multiplexers—bit interleaving and character interleaving. Bit interleaving is normally used in systems which service synchronous data sources, whereas, character interleaving is normally used to service asynchronous data sources.

When interleaving is accomplished on a bit-by-bit basis, the multiplexer takes one bit from each channel adapter and then combines them into a frame for transmission, prefixing the frame with one or more synchronization characters. This process is illustrated in the top portion of Figure 3.3. When interleaving is performed on a character-by-character basis, the multiplexer takes a full character from each channel adapter and places it into the section interval of the message frame as illustrated in the lower portion of Figure 3.3.

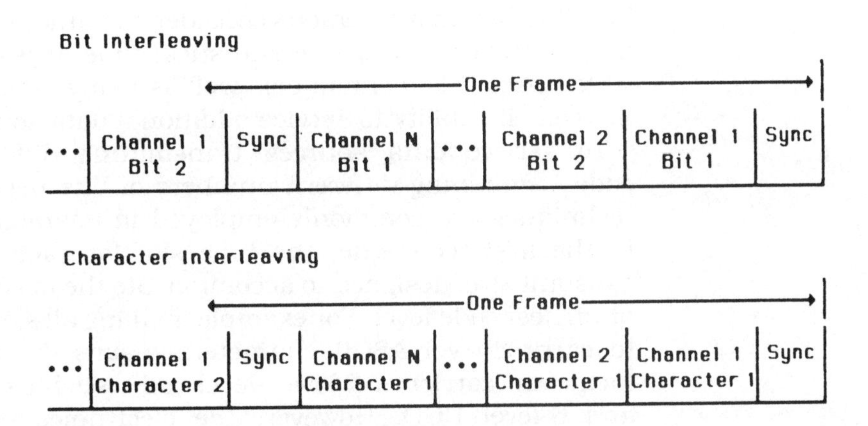

Figure 3.3 TDM multiplexing techniques. When interleaving is accomplished bit-by-bit (top), the first bit from each channel is packed into a frame for transmission. When interleaving is accomplished on a character-by-character basis (bottom), one character from each channel is packed into a frame for transmission.

Since the character-by-character interleaving method preserves all bits of a character in sequence, it became possible for TDM designers to develop circuitry which stripped unnecessary information from characters prior to their placement in the multiplexing interval of a frame. At the opposite end of the data link, the information previously removed from characters in the multiplexing interval was added back to each character by circuitry in the distant multiplexer prior to that device transmitting the demultiplexed data to its channel destination. Normally, information stripping was used with asynchronous data, with the start, stop, and parity bits of characters stripped prior to multiplexing and added back to the character after it was demultiplexed. Examples of asynchronous data sources that could be stripped include the Teletype Model 33 and an IBM Personal Computer transmitting asynchronous data, where each transmitted character contains 10 or 11 bits which include a start bit, 7 or 8 data bits, a parity bit when 7 data bits are used, and 1 or 2 stop bits. When the bit interleaved multiplexing method is used, all 10 or 11 bits would be transmitted to preserve character integrity, whereas, in a character interleaved system, the start and stop bits as well as the parity bit can be stripped from each character. Since this technique increased the efficiency of servicing a 7- or 8-data bit asynchronous character by 3/11ths or 2/10ths, it permitted a few additional low-speed data sources to be serviced by a character interleaved TDM. In fact, many persons consider a character-stripping TDM as the predecessor of the statistical time division multiplexer, which uses this technique as well as many other techniques to increase its ability to service additional data sources.

To service data sources transmitting different character codes containing different numbers of bits per character, two techniques are commonly employed in character interleaving. In the first technique, the time slot for each character is a constant size, designed to accommodate the maximum bit width or highest code level. For example, making all slots large enough to carry 8-level ASCII characters makes the multiplexer an inefficient carrier of a lower-level code, such as 5-level Baudot and 6-level BCD. However, the electronics required in the device and its costs are reduced. This method of TDM is also commonly used in modern microprocessor-based multiplexers as it considerably reduces the complexity of programming. The second technique used in character interleaved TDMs results in the assignment of proportional time slots in which each time slot is sized to the width of each character according to its bit size. This technique maximizes the efficiency of the multiplexer; however, the complexity of fixed logic TDMs and their cost increases. When TDMs are based upon the use of microprocessors, the assignment of proportional time slots increases the complexity of programming which also results in an increase in the cost of the multiplexer.

In comparing bit interleaving to character interleaving, there are advantages and disadvantages associated with each technique. Although bit interleaving equipment is normally less expensive than character interleaving equipment, it is also less efficient when used to service asynchronous terminals. However, bit interleaving multiplexers offer the advantage of faster resynchronization and shorter transmission delay, since character interleaved multiplexers must wait to assemble bits into characters, whereas a bit interleaved multiplexer can transmit each bit as soon as it is received from a data source. To obtain the advantages of bit and character interleaving some multiplexers incorporate both techniques. In doing so synchronous data sources are usually bit interleaved, while asynchronous data sources are character interleaved.

TDM efficiency

The keys to the efficiency of conventional TDMs include the method used to assign data sources to time slots on the high-speed line and the relationship between the scan cycle and the use of a synchronization character.

Most character-based TDMs assign slots to low-speed channels in direct proportion to their character rates and bit size. Under this method a 300 bps data source would be assigned twice as many slots as a 150 bps channel. Thus, in determining the efficiency of a particular TDM you must consider the lowest slot size in comparison to the operating rate of the data sources you plan to multiplex. For example, if the smallest slot size is for 300 bps and your organization has a number of antiquated 134.5 bps terminals that require multiplexing you will have to use 300 bps slots to service each low-speed terminal.

The second area you should consider is the relationship between the scan cycle and the use of a synchronization character appended to the multiplexing frame. Although many TDMs add one synchronization character as a prefix to data gathered from one cycle, some multiplexers insert a synchronization character once for multiple frames. The latter method is more efficient as it significantly reduces the overhead of the multiplexer.

To determine the effect of the synchronization character upon the multiplexing capacity of a TDM, you should first divide the high-speed line rate by eight. This provides the maximum number of characters per second that can be serviced by the high-speed line. Next, determine the frame format of the TDM, including the number of cycles into which a synchronization character is inserted and the operating rate of each slot in a cycle. For example, assume you plan to connect a TDM to a 9600 bps line and can use 1200 bps slots. Then, each cycle could contain a maximum of eight data sources, each operating at 1200 bps. If each cycle requires one synchronization character the overhead would be 1/9, since one character from each channel would be prefixed by a synchronization character. Multiplying the line rate of 9600 by 1/9 indicates that 1066 bps of overhead for synchronization is required. This means that the multiplexer could only multiplex seven 1200-bps data sources. Now suppose the multiplexer inserts one synchronization character every 12 cycles. This means that the synchronization character represents 1/145 of overhead or approximately 67 bps. Although you still could not use this TDM to multiplex eight 1200 bps data sources, you could multiplex seven 1200 bps data sources and three 300 bps data sources or another mixture whose composite input to the TDM is equal to or less than 9300 bps. This is because the synchronization character which represents an actual data rate of 67 bps is assumed to be placed into a 300 bps slot if that is the lowest slot supported by the TDM.

3.3 TDM APPLICATIONS

The basic TDM is primarily used as a mechanism to reduce the number of separate line connections required by a variety of data sources accessing a central computer site. As such, its primary use is in point-to-point communications systems. Although a basic TDM lacks the intelligence associated with other types of TDMs, such as statistical multiplexers, through appropriate cabling it can be used to accomplish some types of networking normally associated with other types of multiplexers. In this section, we will examine several TDM applications, including cabling tricks that enable a TDM to provide some of the switching capability associated with statistical multiplexers.

Point-to-point multiplexing

The most commonly used TDM configuration is the point-to-point system, which is shown in Figure 3.4. This type of system, which is also called a two-point multiplex system, links a mixture of terminals to a centrally located multiplexer. As shown, the terminals can be connected to the multiplexer in a variety of ways. Terminals can be connected by a leased line running from the terminal's location to the multiplexer, by a direct connection if the user's terminal is within the same building as the multiplexer and a cable can be laid to connect the two, or terminals can use the switched network to call the multiplexer over the dial network. For the latter method, since the connection is not permanent, several terminals can share access to one or more multiplexer channels on a contention basis.

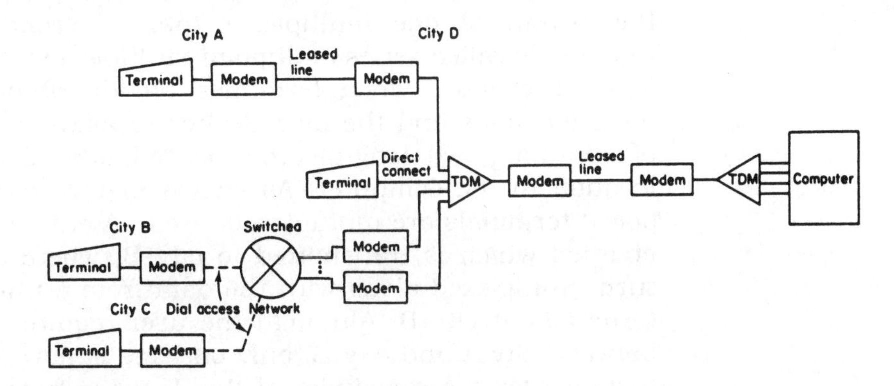

Figure 3.4 Time division multiplexing point-to-point. A point-to-point or two-point multiplexing system links a variety of data users at one or more remote locations to a central computer facility.

As shown in Figure 3.4, the terminals in cities B and C use the dial network to contend for several multiplexer channels which are interfaced to automatic answering modems connected to the dial network. Whenever terminals access those channels they exclude other terminals from access to that particular connection to the system. As an example, you might have a network which contains 50 terminals within a geographical area wherein between 10 to 12 are active at any time, and one method to deal with this environment would be through the installation of a 12-number rotary interfaced to 12 autoanswer modems which in turn are connected to a 12-channel multiplexer. If all of the terminals were located within one city, the only telephone charges that the user would incur in addition to those of the leased line between multiplexers would be local call charges each time a terminal user dialed the local multiplexer number.

Although modems are shown connecting remote terminals to the multiplexer at city D as well as interconnecting TDMs, you can also use digital transmission facilities to access the multiplexer as well as to interconnect multiplexers. When digital transmission facilities are used, the modems will be replaced by the use of channel service units (CSUs) and data service units (DSUs), which are now normally manufactured as a combined CSU/DSU and referred to as DSU.

Series multipoint multiplexing

A number of multiplexing systems can be developed by linking the output of one multiplexer into a second multiplexer. Commonly called series multipoint multiplexing, this technique is most effective when terminals are distributed at two or more locations and the user desires to alleviate the necessity of obtaining two long-distance leased lines from the closer location to the computer. As shown in Figure 3.5, four low-speed terminals are multiplexed at city A onto one high-speed channel which is transmitted to city B, where this line is in turn multiplexed along with the data from a number of other terminals at city B. Although the user requires a leased line between city A and city B, only one line is now required to be installed for the remainder of the distance from city B to the computer at city C. If city A is located 50 miles from city B, and city B is 2000 miles from city C, then 2000 miles of duplicate leased lines are avoided by using this multiplexing technique.

Multipoint multiplexing requires an additional pair of channel cards to be installed at multiplexers 2 and 3 and higher-speed modems or DSUs to be interfaced to those multiplexers to handle the higher aggregate throughput when the traffic of multiplexer 1 is routed through multiplexer 2; but, in most cases the cost savings associated with reducing duplicated leased lines will more than offset the cost of the extra equipment. Since this is a series arrangement a failure of either TDM2 or TDM3 or a failure of the line between these two multiplexers will terminate service to all terminals connected to the system.

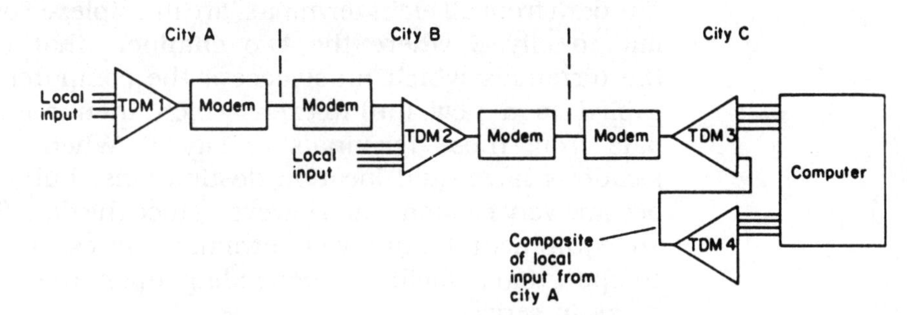

Figure 3.5 Series multipoint multiplexing. This technique is accomplished by connecting the output of one multiplexer as input to a second device.

Hub-bypass multiplexing

A variation of series multipoint multiplexing is hub-bypass multiplexing. To be effectively used, hub-bypass multiplexing can occur when a number of remote locations have the requirement to transmit to two or more locations. To satisfy this requirement, the remote terminal traffic is multiplexed to a central location which is the hub, and the terminals which must communicate with the second location are cabled into another multiplexer which transmits this traffic, bypassing the hub.

Figure 3.6 illustrates one application where hub bypassing might be utilized. In this example, eight terminals at city 3 require a communications link with one of two computers; six terminals always communicate with the computer at city 2, while two terminals use the facilities of the computer at city 1.

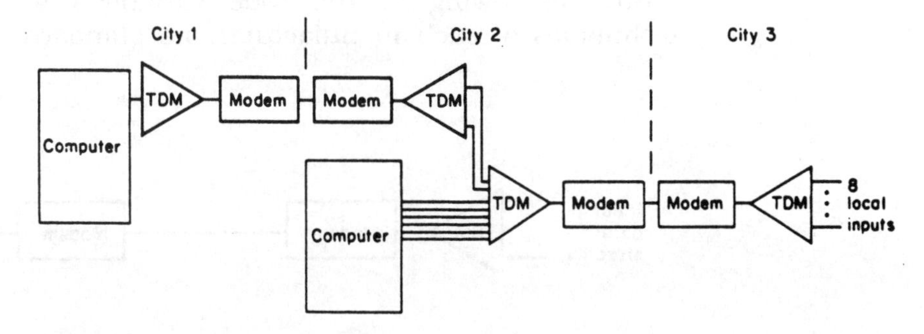

Figure 3.6 Hub-bypass multiplexing. When a number of terminals have the requirement to communicate with more than one location, hub-bypass multiplexing should be considered.

The data from all eight terminals are multiplexed over a common line to city 2 where the two channels that correspond to the terminals which must access the computer at city 1 are cabled to a new multiplexer, which then remultiplexes the data from those terminals to city 1. When many terminal locations have dual location destinations, hub-bypassing can become very economical. However, since the data flows in series, an equipment failure will terminate access to one or more computational facilities, depending upon the location of the break in service.

Although hub-bypass multiplexing can be effectively used to connect collocated terminals to different destinations, if more than two destinations exist a more efficient switching arrangement can be obtained by the employment of a port selector or a statistical multiplexer that has port selection capability. The reader is referred to Chapter 4 for information concerning the port selection capability of statistical multiplexers.

Front-end substitution

Although not commonly utilized, a TDM may be installed as an inexpensive front end for a computer, as shown in Figure 3.7. When used as a front end, only one computer port is then required to service the terminals which are connected to the computer through the TDM. The TDM can be connected at the computer center, or it can be located at a remote site and connected over a leased line and a pair of modems or DSUs. Since demultiplexing is conducted by the computer's software, only one multiplexer is necessary.

However, owing to the wide variations in multiplexing techniques of each manufacturer, no standard software has

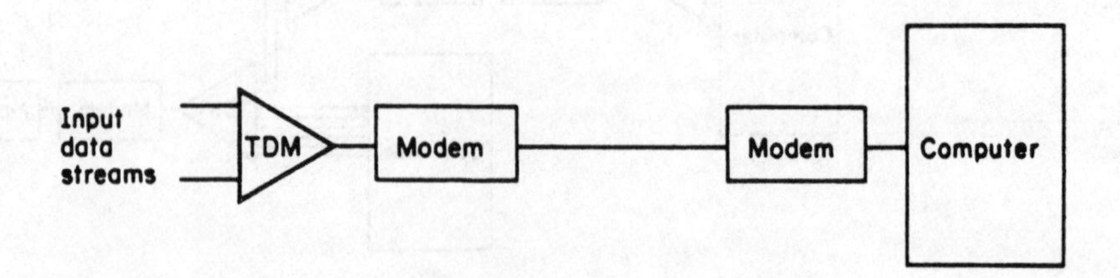

Figure 3.7 TDM system used as a front end. When a TDM is used as a front-end processor, the computer must be programmed to perform demultiplexing.

been written for demultiplexing; and, unless multiple locations can use this technique, the software development costs may exceed the hardware savings associated with this technique. In addition, the software overhead associated costs may exceed the hardware savings associated with this technique. Furthermore, the software overhead associated with the computer performing the demultiplexing may degrade its performance to an appreciable degree and must be considered.

Inverse multiplexing

A specialized type of TDM is the inverse multiplexing system. As shown in Figure 3.8, inverse multiplexing permits a high-speed data stream to be split into two or more slower data streams for transmission over lower-cost lines and modems.

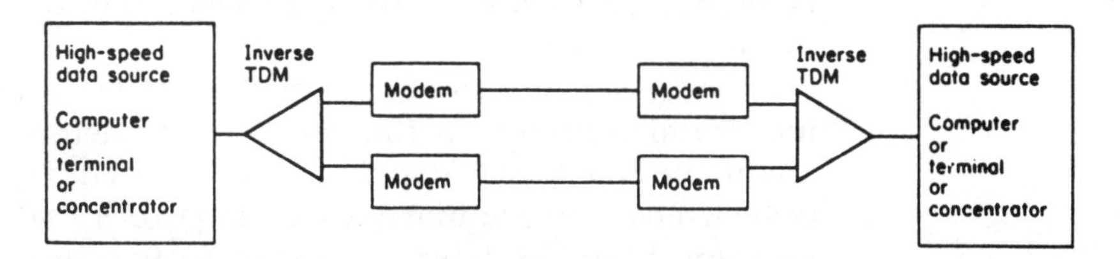

Figure 3.8 Inverse multiplexing. An inverse multiplexer splits a serial data stream into two or more individual data streams for transmission at lower data rates.

Because of the tariff structure associated with wideband facilities, the utilization of inverse multiplexers can result in significant savings in certain situations. As an example, their use could permit 38 400 bps transmission over two voice-grade lines at a fraction of the cost which would be incurred when using analog wideband facilities.

The use of inverse multiplexers has declined in proportion to the growth in the availability of high-speed digital transmission facilities, including 56 kbps DDS and fractional T1 circuits. Since the cost of those digital transmission facilities are usually less than the cost of equivalent analog facilities, most organizations have migrated to digital transmission services that provide a data rate at or beyond that obtainable through the use of two or more voice-grade analog lines and a pair of inverse multiplexers. Although the use of inverse multiplexers peaked during the early 1980s, organizations may wish to consider their use at locations where digital transmission facilities are not available, since the use of inverse multiplexers and two or more voice-grade lines may be the only mechanism to obtain a high-speed data transmission capability.

3.4 MULTIPLEXING ECONOMIES

The primary motive for the use of multiplexers in a network is to reduce the cost of communications. In analyzing the potential of multiplexers, you should first survey terminal users to determine the projected monthly connect time of each terminal. Then, the most economical method of data transmission from each individual terminal to the computer facility can be computed. To do this, direct dial costs should be compared with the cost of a leased line from each terminal to the computer site.

Once the most economical method of transmission for each individual terminal to the computer is determined, this cost should be considered the 'cost to reduce'. The telephone mileage costs from each terminal city location to each other terminal city location should be determined in order to compute and compare the cost of utilizing various techniques, such as line dropping and the multiplexing of data by combining the data streams of several low- to medium-speed terminals' into one high-speed line for transmission to the central site.

In evaluating multiplexing costs, the cost of telephone lines from each terminal location to the 'multiplexer center' must be computed and added to the cost of the multiplexer equipment. Then, the cost of the high-speed line from the multiplexer center to the computer site must be added to produce the total multiplexing cost. If this cost exceeds the cumulative most economical method of transmission for individual terminals to the central site, then multiplexing is not cost-justified. This process should be reiterated by considering each city as a possible multiplexer center to optimize all possible network configurations. In repeating this process, terminals located in certain cities will not justify any calculations to prove or disprove their economic feasibility as multiplexer centers, because of their isolation from other cities in a network.

Networking example

An example of the economics involved in multiplexing is illustrated in Figure 3.9. In this example, assume the volume of terminal traffic from the devices located in cities A and B would result in a dial-up charge of $1540 per month if access to the computer in city G was over the switched network. To compute the dial-up charge you would first estimate the number of calls per day and the call duration for the terminal in each city. Next, you would multiply the number of calls by the call duration to obtain the number of connect minutes per day. If your organization normally operates on a 5-day week, you would then multiply the number of connect minutes per day by 22, since an organization that operates on a 5-day week averages 22 working days per month. As a result of this operation you would obtain the monthly call duration in minutes. From a communications carrier's tariff you would next determine the cost per minute for communications based upon the distance band that the terminal's distance from the compute site falls into (that is, most tariffs have a common cost per minute for a range of distances, such as 20 cents per minute for distances from 1000 to 2000 miles). Then, multiplying the monthly call duration by the cost per minute provides an estimate of the monthly dial-up cost for each terminal.

As an example of the computation of monthly dial network charges, let us assume it was estimated that a terminal would require four calls per day, with an average duration of 50 minutes per call. Thus, the daily call duration would be 200 minutes, while the monthly call duration would be 200×22 or 4400 minutes. If the terminal's distance from the computer falls within a mileage band where the cost per minute of switched network use is 35 cents, then the monthly cost for using the switched network would be 4400×0.35 or $1540.

To compute the cost of a leased line, we would examine the applicable tariff for leased lines and determine the monthly cost of the line based upon the distance from the terminal to the computer site. Suppose the cost of leased lines from city A and city B to city G were determined to be $1300 and $1200 per month, respectively. Since those costs are less than the cost of using the switched network the use of leased lines and their cost would become the method to consider and the 'cost to reduce.' Thus, leased lines are shown connecting the terminals in cities A and B to the computer in city G at the top of Figure 3.9, with the monthly cost of each leased line indicated in parentheses.

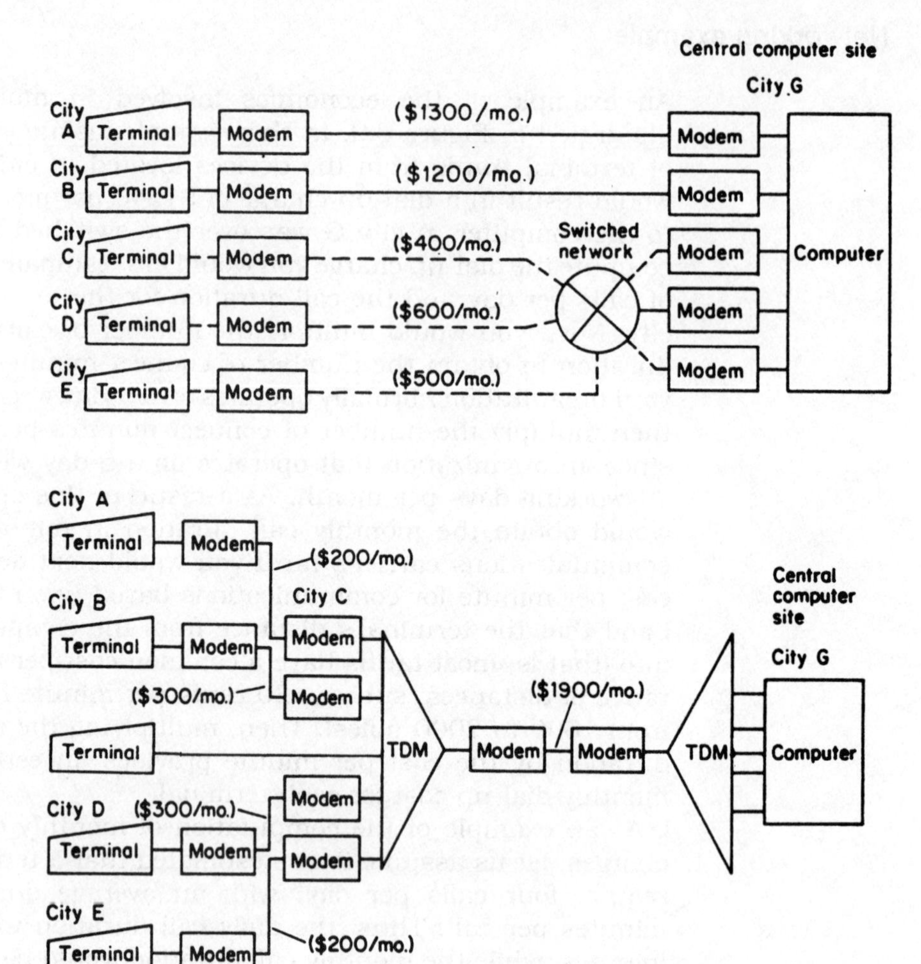

Figure 3.9 Multiplexing economics. On an individual basis, the cost of five terminals accessing a computer system (top) can be much more expensive than when a time division multiplexer is installed (bottom).

Next, let us assume that the terminals at cities C, D, and E only periodically communicate with the computer, and their dial-up costs of $400, $600, and $500 per month, respectively, are much less than the cost of leased lines between those cities and the computer. Then, without multiplexing, the network's most economical communications cost would be:

Location	Cost per month
City A	$1300
City B	$1200
City C	$400
City D	$600
City E	$500
Total cost	$4000

Let us further assume that city C is centrally located with respect to the other cities so we could use it as a homing point or multiplexer center. In this manner, a multiplexer could be installed in city C, and the terminal traffic from the other cities could be routed to that city, as shown in the bottom portion of Figure 3.9. In this example, it was assumed that the most economical method to connect the terminals located in city A and city B with the multiplexer located in city C is via leased lines. Here the monthly cost of each line is $200 and $300 as indicated in parentheses in the lower portion of Figure 3.9. Similarly, the terminals in cities D and E were determined to be more economically connected to the multiplexer in city C via leased lines, with the monthly costs for each line of $300 and $200, respectively. The terminal in city C can be directly cabled to the TDM, hence there is no monthly cost associated with servicing that terminal. However, we must consider the cost of interconnecting the two multiplexers. In this example, we assumed the monthly cost of a leased line connecting the multiplexer in city C to the multiplexer in city G is $1900. In this example, employing multiplexers would reduce the network communications cost to $2900 per month which produces a potential savings of $1100 per month, which should now be reduced by the multiplexer costs to determine net savings. If each multiplexer costs $200 per month, then the network using multiplexers will save the user $700 each month. Exactly how much saving can be realized, if any, through the use of multiplexers depends not only on the types, quantities, and distributions of terminals to be serviced but also on the leased line tariff structure and the type of multiplexer employed.

Alternative configurations

In the previous economic analysis, we examined the use of TDMs to service a mixture of dial-in and directly connected terminals. Of course, most organizations have a mixture of data sources for which multiplexing may be applicable, with a high degree of probability that the communications requirements of one organization will differ from those of the next organization. To assist you in computing the potential economic savings associated with multiplexing let us develop a simple economic model you can tailor to the specific communications requirements of your organization.

Figure 3.10A illustrates non-multiplexed access to a computer from n terminals located at a remote site, while Figure 3.10B illustrates the use of a point-to-point multiplexing system to provide shared access to the computer. In Figure 3.10A we have labeled the communications equipment used to provide terminal access as low-speed modems (LSM); however, they can also represent DSUs if a digital transmission service is used to provide the transmission facility.

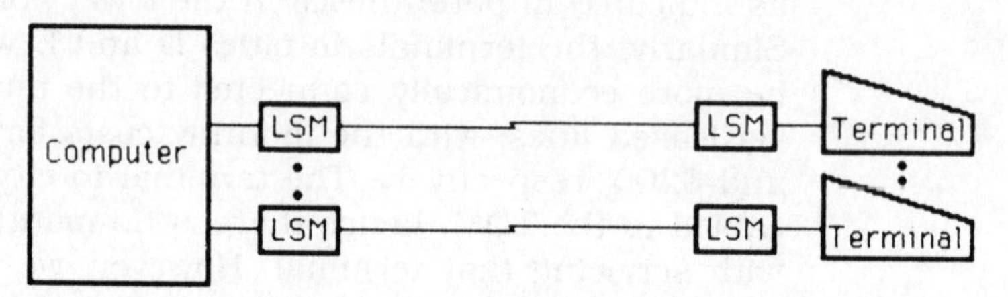

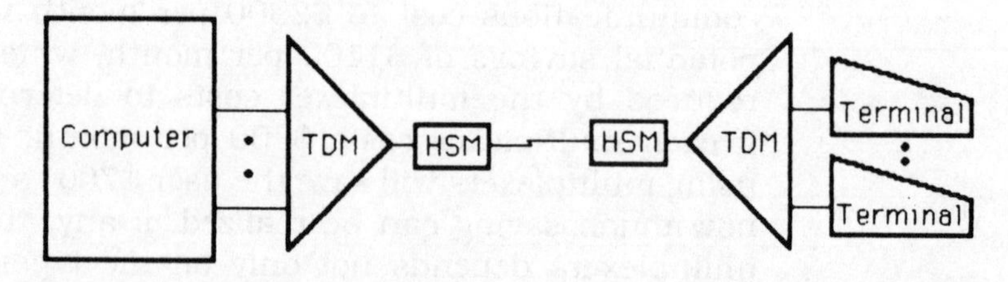

Figure 3.10 Point-to-point multiplexing economics. In point-to-point multiplexing, separate lines and pairs of low-speed modems (LSM) are replaced by a pair of multiplexers (TDM) and a pair of high-speed modems (HSM) connected to a common transmission facility.

One of the problems associated with determining what economic savings, if any, are obtained through multiplexing is the difference in cost measurements between modems, multiplexers, and transmission facilities. Modems and multiplexers, as well as DSUs, are normally purchased, whereas transmission lines are leased on a monthly basis. Thus, one of the first things we must do to perform a viable economic analysis when we are planning to possibly purchase equipment is to determine the useful life of such equipment. In general, most communications equipment is amortized over a 3-year or 36-month period, which represents the typical life of such equipment. Thus, dividing purchase costs by 36 provides a reasonable method to equate one-time expenditures associated with multiplexing to alternative network configurations that do not use multiplexers.

In examining Figure 3.10A, we can define the cost associated with a group of n terminals accessing the computer as follows:

$$C = \frac{LSM}{36} \times n \times 2 + LC \times n$$

where:

n = number of terminals requiring access
LSM = low-speed modem cost
LC = line cost.

Note that LSM is multiplied by n and the result is multiplied by 2 since a low-speed modem is required at each end of the transmission line.

Turning our attention to Figure 3.10B, note that only one transmission line and one pair of high-speed modems (HSM) are now required when multiplexing occurs. Of course, if we are using a digital transmission facility the HSMs would be replaced by a pair of high-speed DSUs. If we denote the cost of multiplexing as CM, we obtain:

$$CM = \frac{2 \times TDM}{36} + \frac{2 \times HSM}{36} + LC$$

where:

TDM = cost of a time division multiplexer
HSM = cost of a high-speed modem
LC = monthly line cost.

Once again, we divided equipment costs by 36, assuming that is the useful life of the TDMs and high-speed modems.

Based upon the above, we would implement multiplexing if $CM < C$. Now suppose our organization has a grouping of terminal devices at different locations. We can first analyze each location as a separate entity to determine if multiplexing is economically viable. Regardless of the outcome, we should also consider using each location as a hub in which data sources at other locations may warrant multiplexing to the hub. In doing so we may find that although the transmission requirements at one location by themselves do not justify multiplexing, when integrated with the requirements from another location multiplexing becomes justified.

Combined FDM–TDM

While FDM equipment used for data multiplexing is limited by the telephone lines' 3 kHz bandwidth, the main limitation on a TDM system is the transmission capability of the high-speed modem or DSU attached to the multiplexer. FDM service, as an example, is usually limited to sixteen 100 bps or eight 150 bps channels, while TDM systems can service a mixture of low- and high-speed data sources whose composite is less than or equal to the speed of the attached modem or DSU. Thus, a single TDM system could service sixty-four 150 bps terminals when interfaced to a 9600 bps modem, whereas eight 8-channel FDM systems might be necessary to provide equivalent service.

Although FDM systems service only low-speed terminals, TDM systems can service a mixture of low- and high-speed data sources, providing the user with more flexibility in both network design and terminal selection. As mentioned previously, while TDM systems favor point-to-point applications, FDM systems are well suited for multidrop configurations where a number of widely separated terminals can be serviced most economically over a single multipoint line. Although terminal quantities, locations, and transmission rates will frequently dictate which type of system to use, a mixture of systems can be considered to provide an optimum solution to a few types of network problems. While FDM equipment is often regarded as obsolete technology because of its limited data speed capacity, its inherent capability of providing a multipoint line connection without requiring poll and select software or addressable terminals should keep this equipment in limited use through the 1990s.

So, in a few networks, TDM systems will be used for transmission at high data rates between two widely separated locations, while FDM can be used in the same system to provide multidrop servicing to terminals which, because of transmission rates and locations, may be best serviced by an FDM system. Such a combined system is shown in Figure 3.11. In developing this system, the terminal requirements were first examined and denoted as follows:

City	Terminal quantity	Terminal speed (bps)	Aggregate terminal speed (bps)
A	4	1200	4800
A	2	2400	4800
B	3	2400	7200
C	1	300	300
D	1	300	300
E	1	300	300
F	1	300	300

In examining the terminal requirements of city A, the total aggregate throughput of the four 1200 bps terminals and the two 2400 bps terminals becomes 9600 bps. Since this exceeds the typical capacity of FDM systems, city A becomes a candidate for TDM. Although the two terminals at city B have data transmission rates beyond the upper range of servicing by an FDM system, further examination of the terminals located in cities C, D, E, and F makes them ideal candidates for FDM. If an FDM system is installed to service those cities, based upon geographical distances, we could then install a TDM at city B which we can use to service the two terminals at that city, as well as servicing the output of the TDM in city A and the FDM system.

Thus, instead of three long-distance lines, city B can act as a homing point for all the multiplexers in the system. Since the aggregate throughput of TDM3 will be 18 000 bps, an extra channel card(s), depending upon manufacturer, may be required to make the system run at a standard 19 200 bps data rate. Although this channel will not be utilized by any terminal, the extra 1200 bps capacity is available for servicing additional terminals at a later date. This extra channel card can be one

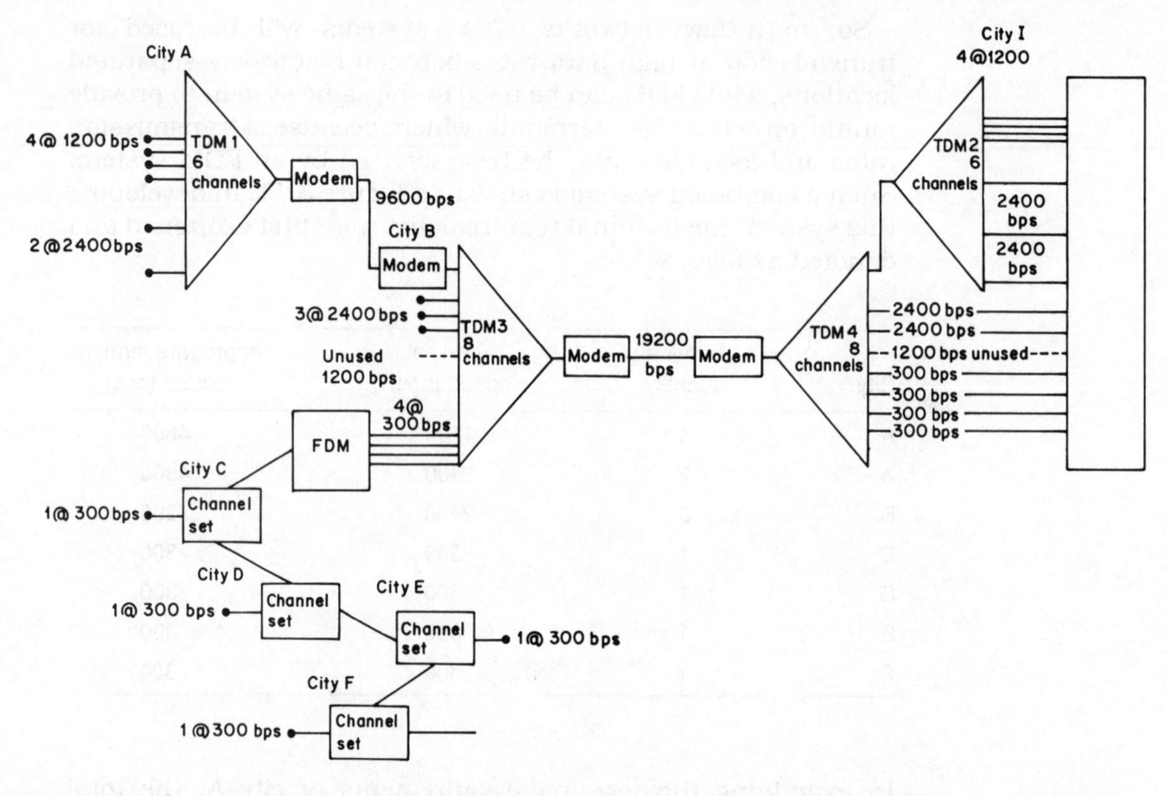

Figure 3.11 Combined FDM–TDM system. Using both FDMs and TDMs in a network permits the capabilities of both devices to be used more advantageously.

1200 bps card (as shown in Figure 3.11) or a number of channel cards whose total capacity adds up to 1200 bps.

At city G only two TDMs are necessary. Since the output of the FDM system is four 300 bps lines which are input to TDM3, TDM4 now demultiplexes this data into four 300 bps channels which are then interfaced to the computer. TDM4, in addition, separates a 9600 bps channel from the 19 200 bps composite speed, and this channel is further demultiplexed by TDM2 into its original six channels which were provided by TDM1. In addition, TDM4 separates the two 2400 bps channels which were multiplexed by TDM3. Thus, the six data channels of TDM1 and the four channels of the FDM system are serviced by the eight-channel TDM3 (which also may include one 1200 bps unused channel for timing). At city G, TDM4 contains eight channels (again, one 1200 bps channel is unused and provides timing) of which the 9600 bps channel is further demultiplexed by the six-channel TDM2 system. Owing to the similarity in channels, TDM4 can be considered the mirror image of TDM3, and TDM2 would be the mirror image of TDM1. With the advent of new families of multiplexers produced by many vendors, only one multiplexer may be required at city G to demultiplex data from all remote sites.

3.5 OTHER TYPES OF TDMs

Although some of us may not recognize it, TDMs are an integral part of two common types of communications equipment used by many organizations—multiport modems and split stream units either built into DSUs or sold as a separate unit for attachment to a DSU. In this section we will examine the operation and utilization of both devices as well as discuss why they will work with most, but not all, data sources.

Multiport modems

The integration of modems and limited-function time division multiplexers into a device known as a multiport modem offers significant benefits to data communications users who require the multiplexing of only a few channels of data. Users who desired to multiplex a few high-speed data channels prior to the introduction of multiport modems were required to obtain both multiplexers and modems as individual units which were then connected to each other to provide the multiplexer and data transmission requirements of the user. Since multiplexers are normally designed to support both asynchronous and synchronous data channels, the cost of the extra circuitry and the additional equipment capacity was an excess burden for many user applications.

The recognition by users and vendors that a more cost-effective, less wasteful method of multiplexing and transmitting a small number of synchronous channels for particular applications led to the development of multiport modems. By the combination of the functions of a time division multiplexer with the functions of a synchronous modem, substantial economies over the past data transmission methods can be achieved for certain applications.

Operation

A multiport modem is basically a high-speed synchronous modem with a built-in TDM that uses the modem's clock for data synchronization, rather than requiring one of its own, as would be necessary when separate modems and multiplexers are combined. In contrast with most traditional TDMs, a multiport modem multiplexes only synchronous data streams, instead of both synchronous and asynchronous data streams. An advantage of the built-in limited-function multiplexer is that it is less complex and expensive, containing only the logic necessary to combine into one data stream information transmitted from as few as two synchronous data channels rather than the minimum capacity of four or eight channels associated with most separate multiplexers. The data channels in a multiport modem normally comprise a number of 2400 bps data streams, with the number of channels available being a function of the channel speed as well as the aggregate throughput of the multiport modem selected by the user.

Selection criteria

When investigating the potential use of multiport modems for a particular application, the user should determine the speed combinations and the number of selectable channels available, as well as the ability to control the carrier function (mode of operation) independently for each of the channels. One 9600 bps multiport modem now being marketed can have as many as six different modes of operation; however, only one mode can be functioning at any given time. As illustrated in Figure 3.12, operating speeds can range in combination from a single channel at 9600 bps through four 2400 bps channels.

Although distinct multiples of 2400 bps are illustrated in Figure 3.12 for simplicity, in actuality, the operating rate of

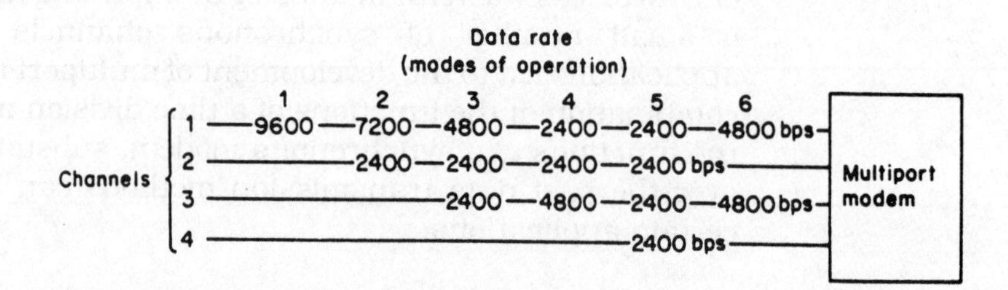

Figure 3.12 9600 multiport modem schematic utilization diagram. Multiport modem with six modes of operation is schematized here to show all possible data rate combinations for networking flexibility.

one channel in each mode will be slightly less than indicated in the illustration. This is because the TDM will use a small portion of the bandwidth of one channel, typically 10 bps, for synchronization during multiple cycles. For example, mode 2 which shows the operation of a 2400 bps and a 7200 bps channel will more than likely operate at 2390 bps and 7200 bps, with 10 bps used for synchronization. Since most modern synchronous data sources, including terminals and computer ports, can adaptively adjust to a variance in a clocking source, they can operate at 2390 bps. However, if your data source cannot adapt to this variance, it cannot be multiplexed on the channel in which a portion of its bandwidth is used for synchronization. In this situation you would want to consider the selection of a different channel to be used for synchronization or the connection of the data source to a different channel. In the remainder of our discussion of multipoint modems and split stream units we will ignore synchronization requirements that slightly lower the operating rate of one channel on the device, but readers should consider the ability of their data sources to operate a few percent underspeed to be effectively multiplexed.

Application example

Using the fifth mode of operation shown in Figure 3.12 with four channels at 2400 bps, a typical application of a 9600 bps multiport modem is illustrated in Figure 3.13. This example shows a pair of four-channel 9600 bps multiport modems servicing two interactive synchronous cathode ray tube (CRT) terminals, a synchronous printer operating at up to 300 characters per second or 2400 bps, and eight low-speed synchronous terminals connected by a traditional TDM. The output of the eight-channel TDM is a 2400 bps synchronous data stream, which is in turn multiplexed by the multiport modem. Here the multiport modem's multiplexer combines the eight asynchronous multiplexed 300 bps channels with the three synchronous unmultiplexed 2400 bps channels into a single multiplexed synchronous data stream. At the central site where the computer is located, the multiport modem at that end splits the 9600 bps stream into four 2400 bps data streams; one data stream is then channeled through another eight-channel, traditional TDM, whose eight output data streams in turn are connected to the computer. The eight-channel TDM takes the

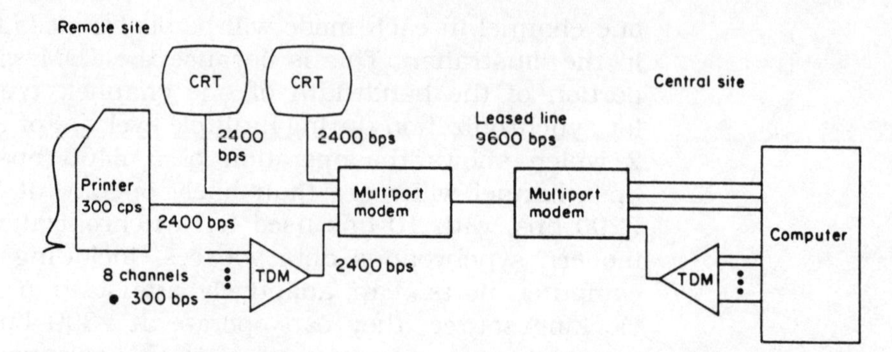

Figure 3.13 Multiport modem application example. A pair of four-channel multiport modems services two CRTs, a 300 cps printer, and eight teletypewriter terminals over a single transmission line.

2400 bps synchronous data stream from the multiport modem and demultiplexes it into eight 300 bps asynchronous data streams, which are passed to the appropriate computer ports. The remaining data streams produced by the demultiplexer in the multiport modem are connected to three additional computer ports. As this example demonstrates, the utilization of a high-speed multiport modem in conjunction with other communications components permits a wide degree of flexibility in the design of a data communications network.

Multiport modem and channel combinations available to the user are listed in Table 3.1 for modems whose data transfer rates range from 19 200 bps to 4800 bps. It should be noted that not all multiport modem channel combinations listed in Table 3.1 may be available for a particular vendor's modem that operates at the indicated aggregate throughput, since some vendors only offer a four-port multiplexer with their modem, while other vendors may offer a six-port or eight-port multiplexer with their device.

Although most manufacturers of multiport modems produce equipment that appears to be functionally equivalent, the system designer should exercise care in selecting equipment because of the differences that exist between modems but are hard to ascertain from vendor literature.

For modem aggregate throughput above 4800 bps, the modes of operation available for utilization by the system designer are quite similar regardless of manufacturer, with the major difference being the number of ports or data channels supported by the multiplexer.

Table 3.1 Multiport modem channel combinations. The wide degree of flexibility that can be provided by multiport modems in a network configuration is a function of several factors: throughput, available modes and channels, and data rates

Modem agregate throughput	Operating mode	Multiport speed combinations 1	2	3	4	5	6	7	8
19 200	1	19 200							
	2	16 800	2400						
	3	14 400	4800						
	4	14 400	2400	2400					
	5	12 000	7200						
	6	12 000	4800	2400					
	7	12 000	2400	2400	2400				
	8	9600	9600						
	9	9600	4800	4800					
	10	9600	4800	2400	2400				
	11	7200	7200	4800					
	12	7200	4800	2400	2400	2400			
	13	7200	2400	2400	2400	2400	2400	2400	
	14	4800	4800	4800	4800				
	15	4800	4800	4800	2400	2400			
	16	4800	4800	2400	2400	2400	2400		
	17	4800	2400	2400	2400	2400	2400	2400	
	18	2400	2400	2400	2400	2400	2400	2400	2400
16 800	1	16 800							
	2	14 400	2400						
	3	12 000	2400	2400					
	4	9600	2400	2400	2400				
	5	9600	4800	2400					
	6	7200	2400	2400	2400	2400			
	7	7200	4800	2400	2400				
	8	7200	7200	2400					
	9	4800	4800	2400	2400	2400			
	10	4800	2400	2400	2400	2400	2400		
14 400	1	14 400							
	2	12 000	2400						
	3	9600	4800						
	4	9600	2400	2400					
	5	7200	7200						
	6	7200	4800	2400					
	7	7200	2400	2400	2400				
	8	4800	4800	4800					
	9	4800	4800	2400	2400				
	10	2400	2400	2400	2400	2400	2400		
12 000	1	12 000							
	2	9600	2400						
	3	7200	4800						
	4	7200	2400	2400					
	5	4800	4800	2400					
	6	4800	2400	2400	2400				
	7	2400	2400	2400	2400	2400			
9600	1	9600							
	2	7200	2400						
	3	4800	2400	2400					
	4	2400	2400	4800					
	5	2400	2400	2400	2400				
	6	4800	4800						
7200	1	7200							
	2	4800	2400						
	3	2400	2400	2400					
4800	1	4800							
	2	2400	2400						

However, at 4800 bps wide variances exist between equipment manufactured by different vendors. While most multiport modem manufacturers offer two-mode capability (one 4800 bps channel or two 2400 bps channels), some manufacturers have a built-in TDM which has the capability of servicing 1200 bps data streams. Another variance concerns the use of an independently controlled carrier signal in some multiport modems. By using multiport modems that permit an independently controlled carrier signal for each channel, data communications users can combine several polled circuits and further reduce leased line charges.

Standard and optional features

A wide range of standard and optional data communications features are available for users of multiport modems, including almost all of the features available in regular non-multiplexing modems; also available are unique multiport modem features such as multiport configuration selection, individual port testing, individual port display, and a data communications equipment (DCE) interface.

The multiport selector feature permits the user to alter the multiport configuration simply by throwing a switch into a new position or pressing a button on the control panel of the modem. This feature can be especially useful for an installation such as the one shown in Figure 3.14 (top), where daytime operations require the servicing of a large number of interactive time-sharing users; while operations at night (Figure 3.14 (bottom)) require the servicing of only two high-speed remote batch terminals. During daytime operations, sixteen low-speed asynchronous, 300 bps terminals with a composite speed of 4800 bps are serviced at this installation by one channel of the multiport modem. One remote batch terminal and a CRT are serviced by two additional channels, each of which operates at 2400 bps. Because of daytime load requirements, the second batch terminal cannot be operated since the modem's maximum aggregate speed of 9600 bps has been reached. On the assumption that the installation does not require the servicing of interactive time-sharing users at night, one possible reconfiguration is shown in Figure 3.14 (bottom). The multiport selector permits both remote batch terminals to be serviced until the start of the next business day by two 4800 bps channels while everything else in this network is shut down.

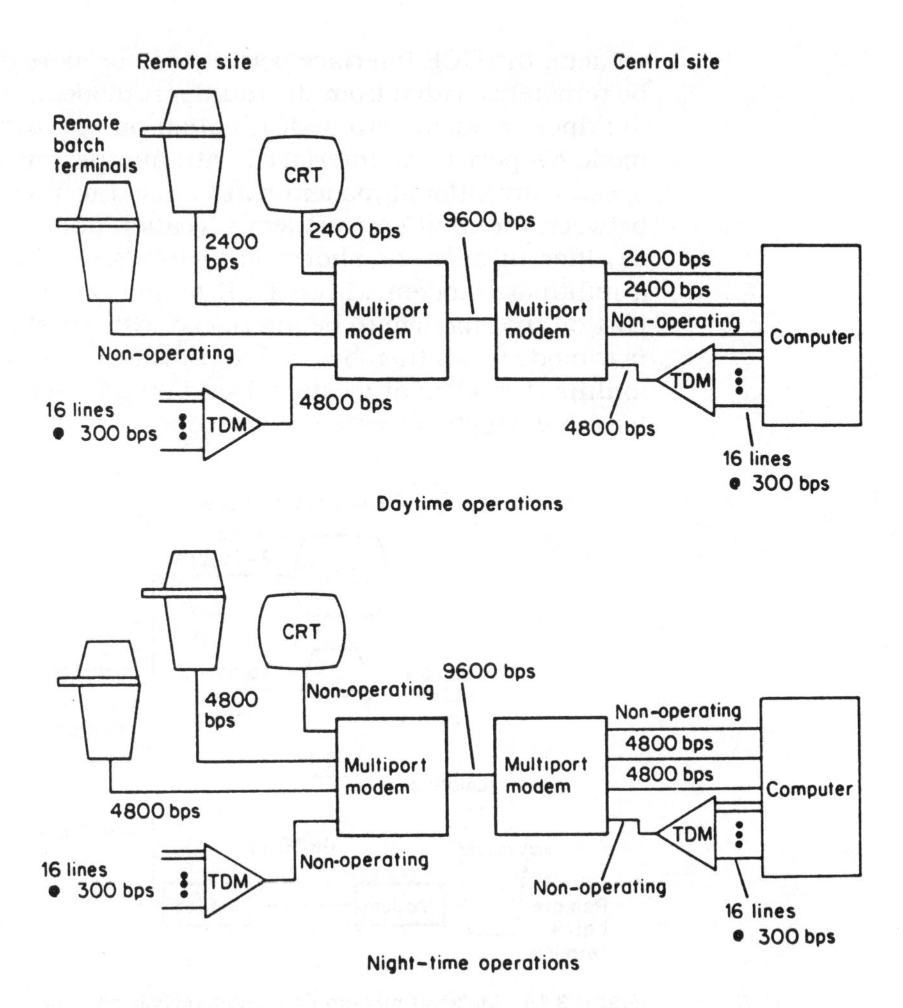

Figure 3.14 Using multiport selector switches. Day (top) and night (bottom) configurations for networks with multiport modems can be varied according to the requirements of different operations.

Numerous multiport modems contain a built-in test pattern generator and an error detector which permit users of such modems to determine if the device is faulty without the need of an external bit error rate tester. The use of this feature normally permits the individual ports of the modem to be tested.

Another option offered by some multiport modem manufacturers is a data communications equipment (DCE) interface. This option can be used to integrate remotely located terminals into a multiport modem network. Whereas the standard data terminal equipment (DTE) interface may require data sources to be collocated and within a 50 ft radius of the multiport modem, the DCE interface permits one or more data sources to be remotely located from the multiport modem. Installation of a multiport modem with a DCE option on one port permits that modem's port to be interfaced with another modem. This low-speed conventional modem can be used to provide a new link between the multiport modem's location and terminals located at different sites. As shown in Figure 3.15, the installation of a multiport modem with a DCE option on port 3 permits the port on that modem to be interfaced with another modem. This new modem can then be used to provide a new link between the multiport modem at location 1 and an additional remote batch terminal which is located at a second site.

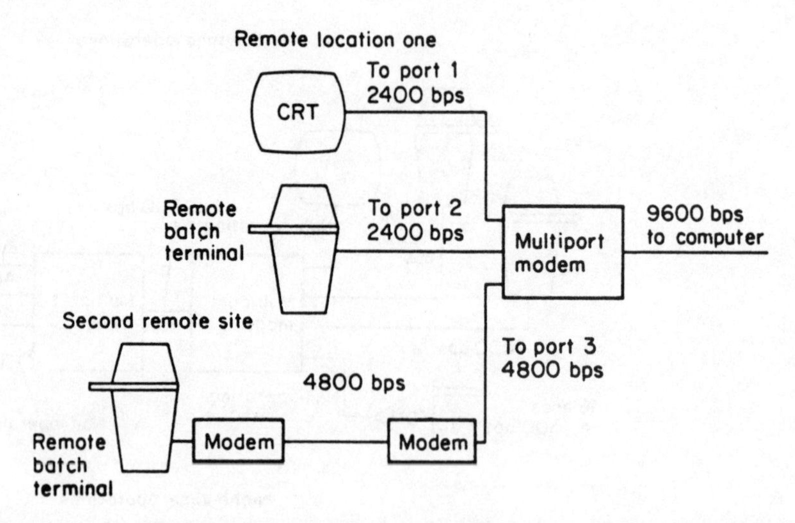

Figure 3.15 Multiport modem data communications equipment option. Using a data communications equipment (DCE) interface on port 3 of the multiport modem permits a second remote site to share the communications line from the first site to the computer.

Split stream DSUs

Similar to the rationale for incorporating TDMs into modems, several vendors have built limited-functioning time division multiplexers into DSUs. The resulting device, commonly referred to as a split stream DSU, provides an economical method to share the use of a digital transmission facility by several synchronous data sources.

Most split stream DSUs marketed are presently limited to two- or four-port units. Similar to the TDMs used in multipoint modems, a portion of the bandwidth on one channel is used for synchronization which slightly reduces the available data transmission rate on that channel. The channel with the reduced data rate must be set to provide clocking to a collocated data source and that data source must be capable of adjusting to a non-standard data rate.

Applications

Figure 3.16 illustrates the use of a stand-alone split stream unit interfaced to a DSU. In this example it was assumed that three ports of a four-port unit were used, with the collocated terminal serviced by a non-standard clocking source. Here the split stream unit is similar to the TDM incorporated into a multipoint modem in that the user can select various combinations of data transfer rates on up to four individual channels, up to the maximum capability of the digital transmission line. The split stream unit plugs directly into the DSU, with operational settings at one-half or one-quarter (except for one channel operating typically 10 bps less) of the specified digital line operating rate. In the example illustrated the multiplexer enables data sources at several locations to include an off-net

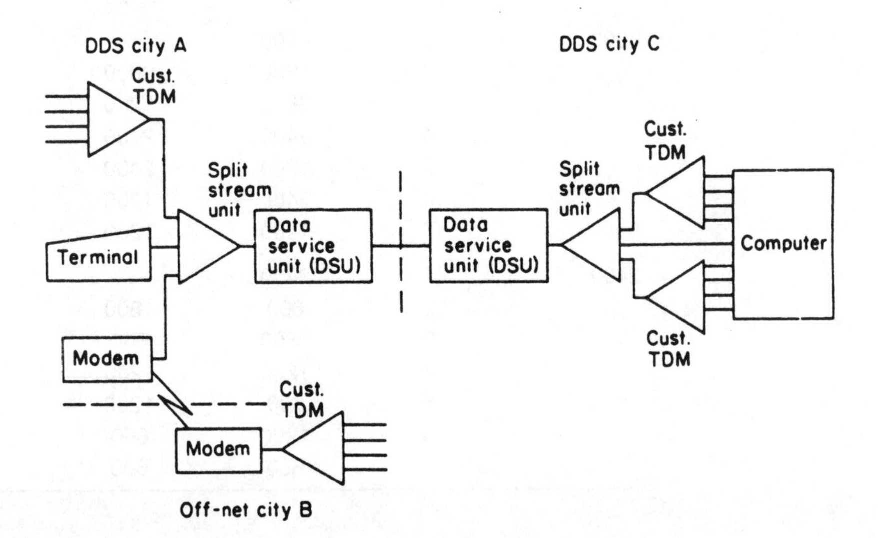

Figure 3.16 Multiplexing over DDS utilizing split stream units. An inexpensive split stream unit, or limited-function synchronous multiplexer, can offer considerable flexibility when interfacing into the DDS network.

city where digital transmission is not available to be multiplexed at a hub location onto a long distance digital transmission facility.

To illustrate the potential networking capability afforded by split stream units, Table 3.2 lists the operational modes of a four-port device designed for use on 2400, 4800, and 9600 bps Dataphone Digital Service (DDS) digital transmission facilities. In comparing the entries in Table 3.2 to the entries in Table 3.1, readers will note that the operating capabilities of split stream units are equivalent to those of four-port multiport modems. With the recent introduction of 19.2 kbps DDS service, it is probably a reasonable expectation that vendors will introduce six- and eight-port split stream units in the near future.

Table 3.2 Split stream unit modes of operation

DDS rate	Mode	SSU	Channel	Data	Rate
		1	2	3	4
9600	1	9600			
	2	2400	7200		
	3	7200	2400		
	4	4800	4800		
	5	2400	4800	2400	
	6	4800	2400	2400	
	7	2400	2400	2400	2400
4800	1	4800			
	2	1200	3600		
	3	3600	1200		
	4	2400	2400		
	5	1200	2400	1200	
	6	2400	1200	1200	
	7	1200	1200	1200	1200
2400	1	2400			
	2	600	1800		
	3	1800	600		
	4	1200	1200		
	5	600	1200	600	
	6	1200	600	600	
	7	600	600	600	600

4 STATISTICAL MULTIPLEXERS

The focus of this chapter is upon a multiplexing technique incorporated into a new class of multiplexers that can be considered as an evolution of traditional time division multiplexing. The multiplexing process examined in this chapter is commonly referred to as statistical multiplexing, while the resulting equipment is referred to as a statistical multiplexer.

Since traditional TDMs form the basis for the development of statistical time division multiplexers (STDMs), we will first focus our attention upon the TDM process which provides a basis for comparing the operation and efficiency of STDMs to TDMs. After doing so, we will then examine the key operational features of STDMs, including the methods by which their multiplexing frames are constructed, how they control their buffer occupancy through the process of flow control, and methods by which the data handling capacity of STDMs can be compared to TDMs. This will be followed by an in-depth examination of the reporting statistics produced by this class of multiplexing equipment, their features, and typical applications suitable for the use of STDMs.

4.1 COMPARISON TO TDMs

In a traditional time division multiplexer, data streams are combined from a number of devices onto a single path so that each data source has a time slot assigned for its exclusive use. This method of multiplexing is inefficient, since a time slot is reserved for each connected device, regardless of the state of activity of the device. When a device is inactive, the

TDM pads the slot with nulls and cannot use the slot for other purposes. These nulls are inserted into the message frame, since demultiplexing occurs by the position of characters in the frame. Thus, if the use of nulls to represent the inactivity of devices is eliminated, a scheme must be used to indicate the origination port or channel of data. Otherwise, there would be no way to correctly reconstruct the data during the demultiplexing process.

The key to the greater efficiency of a statistical multiplexer over a traditional TDM is the dynamic allocation of time slots by the STDM. This results in a more efficient utilization of the high-speed transmission medium, permitting the STDM to service more terminals than a traditional time division multiplexer. This technique of allocating time slots on a demand basis is known as statistical multiplexing and means that data is transmitted by the multiplexer only from devices that are active. Other techniques incorporated into STDMs to increase their efficiency include the stripping of start, stop, and parity bits from asynchronous data and the use of data compression.

Another reason for the greater efficiency of STDMs over TDMs is the ability of the statistical multiplexer to take advantage of the fact that most data sources employ a half-duplex method of transmission. That is, at any particular time you can expect half of the devices connected to a multiplexer to be transmitting, while the remaining devices will be receiving data. This means that, by taking advantage of the half-duplex nature of transmission, an STDM on average can support twice as many devices as a TDM. For example, a TDM might support eight 1200 bps data sources on a 9600 bps channel. In comparison, an STDM by taking advantage of the nature of most devices operating half-duplex could support 16 devices operating at 1200 bps.

TDM message frame

The major difference between traditional TDMs and STDMs is best illustrated by the construction technique used by each device to build a message frame. Traditional TDMs employ a fixed frame approach as illustrated in Figure 4.1. Here, each frame consists of one character or one bit for each input channel scanned during a particular period of time. As illustrated, even when a particular terminal is inactive, the slot assigned to that device is included in the transmitted message frame since the

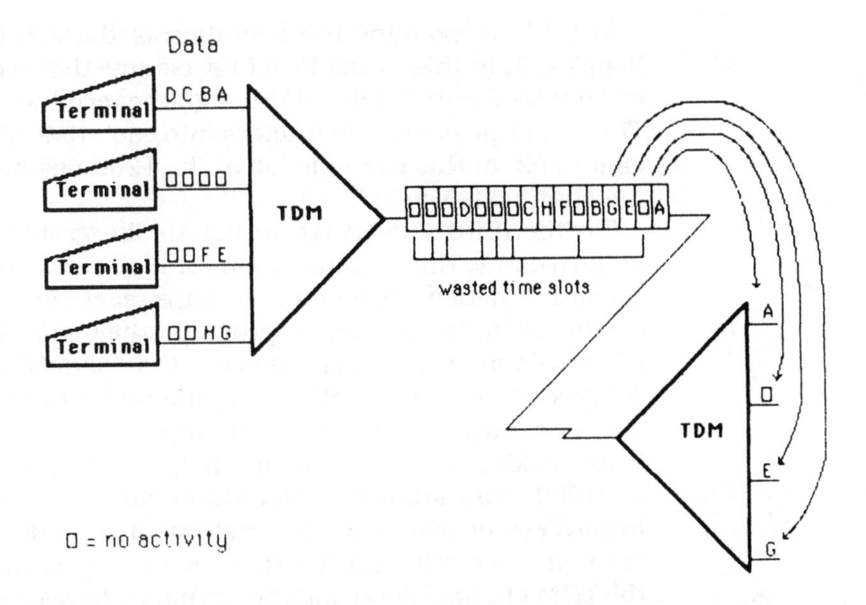

Figure 4.1 Traditional TDM frame construction. Traditional TDMs contain a time slot for each scanned input data source regardless of the activity status of the data source. Since demultiplexing occurs by the position of data in a frame, periods of inactivity are converted to null characters which are inserted into time slots to maintain positioning of data.

presence of a null character in the time slot is required to correctly demultiplex the data. In the right hand portion of Figure 4.1 the demultiplexing process which is accomplished by time slot position is illustrated.

If we consider the typical activity of terminal users we can observe why a TDM is inefficient. Let us assume four terminals, each operating at 1200 bps, are connected to a TDM, as illustrated on the left of Figure 4.1. If the first terminal user is entering data, what is the actual data rate of the terminal that is transmitting at 1200 bps? Assuming a very fast typist entering data at 50 to 60 words per minute, if each word averages six characters, the typist's maximum actual data rate is 60 words per minute × 6 characters per word × 8 bits per character divided by 60 seconds per minute, or 48 bits per second! This means that there are actually numerous time slots on channel 1 that are filled with nulls, as the typist cannot enter data fast enough to assign each character typed at the terminal to contiguous time slots reserved for channel 1.

Now let us examine the operation of the second terminal in Figure 4.1. In this example, let us assume the terminal operator has left for a cup of coffee. Although the terminal is inactive, the TDM must place null characters into each time slot reserved for channel 2. In this example, all of the 1200 bps operating rate of the channel is wasted.

For the third and fourth terminals illustrated in Figure 4.1, let us assume the terminal operators are performing a common human function—thinking. Perhaps each operator issued a command or a line for a program they are developing and received that proverbial message that something is wrong—'Syntax Error.' Thus, after entering some data and receiving an error message the terminal user is thumbing through a book, asking an associate for help or simply staring at the terminal in an attempt to determine how to correct a problem. Regardless of what the terminal operator is doing, when they are doing something other than using the terminal, activity of the TDM channel servicing the terminal changes from time slots partially filled with data to time slots filled with nulls. Only when transferring a file, listing the contents of a file residing on a computer connected to a TDM at the opposite end of the circuit, or directing system output to a printer attached to a terminal, will the time slots reserved for use by a specific channel be filled with data characters. However, since the probability of such activities is small with respect to the typical use of a terminal, we can reasonably expect nulls to account for a large percentage of the contents of time slots, probably exceeding 75% of the contents of all time slots for substantial periods of time during the day and well over 99% of all time slots in the evening. Thus, the TDM technique contains obvious inefficiencies and has resulted in the development of statistical time division multiplexers.

Statistical frame construction

A statistical multiplexer employs a variable frame building technique which takes advantage of terminal idle times to enable more terminals to share access to a common circuit. The use of variable frame construction technology permits previously wasted time slots to be eliminated, since control information is transmitted with each frame to indicate which terminals are active and have data contained in the message frame.

Address and byte count method

One popular statistical multiplexing technique involves buffering data from each data source and then transmitting the data with an address and byte count. The address is used by the demultiplexer to route the data to the correct channel, while the byte count indicates the quantity of data to be routed to that channel. Figure 4.2 illustrates the message frame of a four-channel statistical multiplexer employing the address and byte count frame composition method during a certain time interval. In this example, the multiplexer uses a frame structure similar to HDLC to include the use of a cyclic redundancy check (CRC) for error detection. Later in this chapter we will examine how STDMs provide data integrity. Since channels 3 and 4 were assumed to be inactive during the defined time interval, there was no address and byte count, nor data, from those channels transmitted on the common circuit.

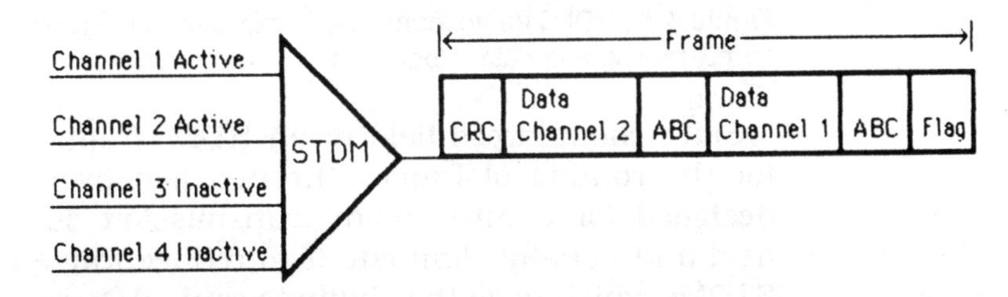

Figure 4.2 STDM address and byte count frame construction. In a statistical multiplexer, data from active channels is first buffered until a predefined number of characters is received. Then the channel data is prefixed by an address and byte count (ABC) and used as an entity to construct the message frame.

In examining Figure 4.2, the reader will note that the data from each channel is of variable length. Some statistical multiplexers that use an address and byte count method actually buffer a small quantity of data prior to inserting the data into a variable frame. Such multiplexers use one or more techniques which govern the maximum quantity of buffered data. For example, some STDMs are designed to wait until either 32 characters or a carriage return is encountered or a predefined period of time occurs prior to forming the address and byte count and forwarding the buffered data. For asynchronous transmission the multiplexer can be set to generate a local echo which causes characters typed at a terminal to be immediately echoed back to the terminal. Thus, although characters are buffered and do not immediately flow onto the high-speed line, for all practical purposes the terminal user is unconscious of this fact. The reason 32 characters was selected as a decision criteria is that it represents the average line length in an interactive transmission session.

To increase the efficiency of STDMs, vendors have developed a variety of frame construction techniques. Most of these techniques are based upon the use of a variable frame which is either compatible with the CCITT Higher Level Data Link Control frame structure or resembles that frame structure. Figure 4.3 illustrates the general format of the HDLC frame structure.

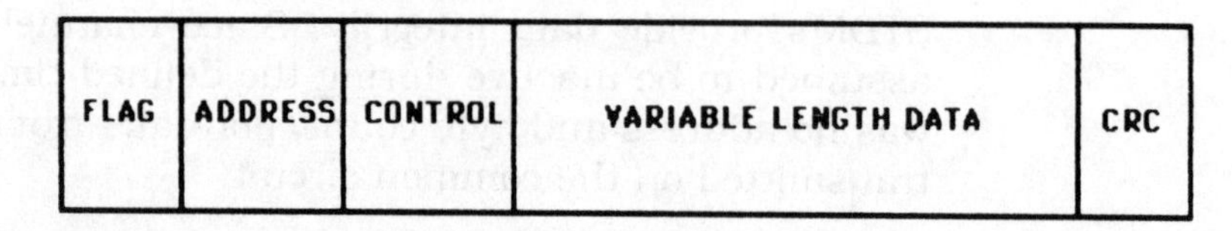

Figure 4.3 HDLC frame format. By encapsulating multiplexed data into an HDLC frame, routing and data integrity can be supported.

Since the address field in an HDLC frame is primarily used for the routing of frames through a network, simple STDMs designed for point-to-point transmission do not require that field and usually eliminate it to reduce frame overhead. Other STDMs which have the ability to route data over different paths interconnecting STDMs, and which some manufacturers refer to as intelligent nodal processors, require the use of an address field to route data. Equipment in this category incorporates an address field in the HDLC frame which provides the mechanism for routing data through a series of STDMs to its ultimate destination.

Within a manufacturer's product line you may find some STDMs that cannot be internetworked with other STDMs. The primary reason for this lack of interoperability is the absence of an address field and software which operates upon entries in the address field in low-end STDMs, making those multiplexers incompatible with a vendor's high-end products.

Dual addressing format

One interesting variation to the previously described address and byte count method of frame composition involves a dual

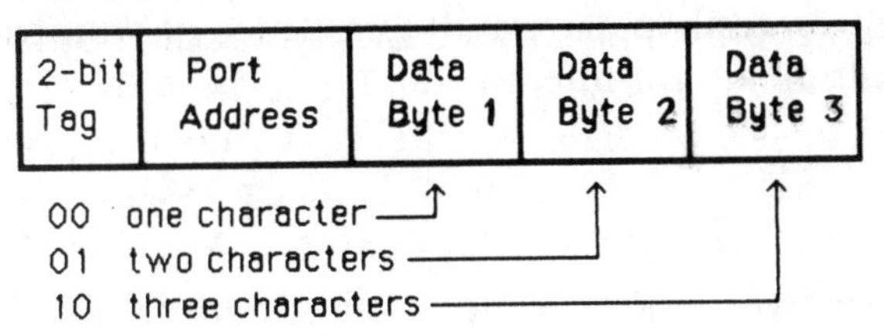

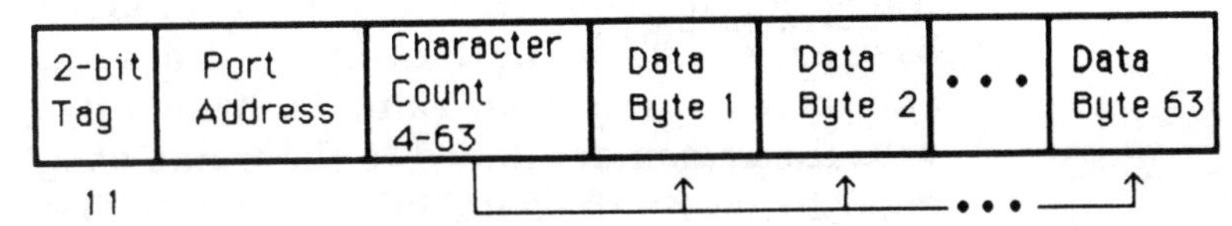

Figure 4.4 Dual data structure format. Under this dual data structure format, the use of a 6-bit port address limits the number of data channels to 64 regardless of the data rate or activity of each channel.

addressing format which is encapsulated into a variable length data field in the HDLC frame. Figure 4.4 illustrates the dual addressing format in which one 8-bit byte is used both to identify the format and the port address of the following data.

At the top of Figure 4.4, format A is identified by bit compositions 00, 01, and 10 in the first two positions of the first byte. The remaining six bits are used to identify the port address of the following data, which limits this addressing scheme to 64 port addresses (0 to 63) and explains why STDMs using this technique are constrained to servicing 64 data sources regardless of the operating rates of the data sources. One interesting aspect of this format is that the 2-bit tag at the beginning of the first byte also identifies the number of data bytes in the field. Here the 2-bit value plus one indicates the number of data bytes.

Under the second format which is identified by the 2-bit tag 11, a count field of six bits follows the 6-bit port address field. The count field is used to indicate the occurrence of 4 to 63 characters that follow the count field.

Through the use of two formats, the extra overhead associated with a count field is eliminated for the transmission of groups of three or less characters. In addition, an STDM that uses this type of dual format can reduce buffering delays as it can more rapidly place small quantities of data from a port onto the multiplexing frame.

Data integrity

Both the address and byte count, as well as the dual format technique and other frame building methods, are based upon the use of a frame structure which includes a CRC. This structure is a requirement for statistical time division multiplexing as it provides a mechanism for determining if a transmission error has occurred. That is, the CRC is computed by treating all of the bits in the frame less the flag bits as one binary value. That value is divided by a fixed polynomial, with the remainder added to the frame as a 2-byte CRC. At the opposite end of the data link, another STDM performs the same operation on the received data, generating a 'local' CRC. The local CRC is then compared to the transmitted CRC. If both match, the frame is considered to have been received without error. If the CRCs do not match, an error is assumed to have occurred and the receiving STDM will request the transmitting STDM to retransmit the frame.

Instead of holding up transmission to process the retransmission request, most STDMs use the control field to allow up to seven transmit and receive frames to be unacknowledged and support selective retransmission. This allows one multiplexer to continue transmission and insert the previously rejected frame into the data stream. Of course, the receiving STDM must wait until the previously rejected frame is received correctly so it can demultiplex data in its proper sequence.

Operational problems

Although statistical multiplexers are more efficient than traditional TDMs with respect to the servicing of data sources, there are a few potential drawbacks associated with their use. STDM problems include the delays associated with data blocking and queuing when a large number of connected terminals become active or when a few terminals transmit large bursts of data. For either situation, the aggregate throughput of the multiplexer's input active data exceeds the capacity of the common high-speed line, causing data to be placed into buffer storage.

Another reason for delays is when a circuit error causes one or more retransmissions of message frame data to occur. Since connected terminals and other devices may continue to send data during the multiplexer-to-multiplexer retransmission cycle, this can also cause the multiplexer's buffer to fill, resulting in additional time delays.

If too much data continues to enter the STDM, its buffer would eventually overflow, causing data to be lost. To prevent buffer overflow, all statistical multiplexers employ some type of control signaling to inhibit data traffic once their buffer is filled to a predefined level. These buffer control techniques inhibit additional transmission into the multiplexer until its buffer is emptied to another predefined level. Once this level is reached, a second control signal is issued to permit data sources to resume their transmission into the multiplexer.

4.2 BUFFER CONTROL

Techniques developed to prevent data input to the buffer memory of statistical multiplexers from overflowing result in the flow control of data. Thus, these techniques are also known as flow control. The three major flow control techniques used by statistical multiplexers include inband signaling, outband signaling, and clock rate adjustment.

Inband and outband signaling

Inband signaling involves the transmission of XOFF and XON characters by the multiplexer to inhibit and enable the transmission of data from terminals and computer ports that recognize these flow control characters. Since many terminals and computer ports do not recognize these control characters, a second common flow control method involves raising and lowering the clear to send (CTS) control signal on the RS-232/CCITT V.24 interface. Since this method of buffer control is outside the data path where data is transmitted on pin 2, this technique of buffer control is also known as outband signaling.

Clocking signal adjustment

Both inband and outband signaling are used to control the data flow of asynchronous devices. Since synchronous devices transmit data formed into blocks or frames, inband or outband signaling could result in the inadvertent subdivision of a block or frame. This could result in the receiving device timing out and negatively acknowledging the transmitter, reducing the throughput of synchronous data sources whenever the activity of the statistical multiplexer reaches a predefined level. To alleviate this problem, many STDM vendors use a technique which involves altering or adjusting the clocking signal to synchronous devices. Thus, a synchronous data source operating at 4800 bps might first be reduced to operating at 2400 bps by the multiplexer halving the clock rate. Then, if the buffer in the multiplexer continues to fill, the clock rate might be further reduced to 1200 bps.

Flow control variations

In addition to the three previously discussed methods of flow control, some vendors have added two interesting variations to inband and outband signaling. One inband flow ontrol addition is the ability to program a flow control sequence for terminal equipment that does not recognize the XOFF/XON sequence. This capability could be used for flow controlling many types of Hewlett-Packard equipment which supports the ENQ-ACK sequence. Concerning outband flow control, a few STDMs include the capability to toggle the voltage on pin 22 for equipment that uses that pin on the RS-232 interface for flow control.

Although our prior discussion of flow control was restricted to the STDM controlling the flow of data from external equipment, it is equally important for external equipment to control the flow of data from the statistical multiplexer. For example, consider a personal computer connected via the use of STDMs to a mainframe computer. Assume the personal computer operator is using software that directs data from the mainframe to the PC's printer and the printer is in its near-letter quality (NLQ) print mode of operation. Under this scenario it is highly probable that the print rate is significantly below the data rate at which the personal computer receives data from the mainframe. To prevent the printer's buffer from overflowing and losing data, the printer will flow control the PC. The PC's software in turn will flow control the communications line, which in this example is a channel on the STDM. Thus, the STDM must be capable of recognizing an external flow control signal. In addition, the

STDM must rapidly pass this signal to the STDM at the opposite end of the link, which will then pass the flow control signal to the mainframe's computer port.

Table 4.1 lists the flow control options you should consider for both the STDM controlling external equipment and external equipment controlling a statistical multiplexer. Although all STDMs now support at least two methods of flow control, most statistical time division multiplexers do not support all of the methods of flow control listed in Table 4.1. Thus, readers should carefully consider both the methods of flow control supported by an STDM under consideration as well as the methods of flow control supported by external devices that will be connected to the STDM.

Table 4.1 Flow control methods

STDM controls external equipment		
Type of signaling	Method	Operation
Inband	XON/XOFF	Transmit XON/XOFF sequence
Inband	Programmable	Transmit user selected sequence
Outband	Clear to send	Raise and lower CTS signal
Outband	Ring Indicator	Raise and lower Pin 22 signal
Clocking	Adjust clock rate	Double/halve clock rate
External equipment controls STDM		
Type of signaling	Method	Operation
Inband	XON/XOFF	Respond to XON/XOFF sequence
Inband	Programmable	Respond to user selected sequence
Outband	Clear to send	Respond to high/low CTS signal
Outband	Busy	Respond to high/low pin 25 signal
Clocking	Adjust clock rate	Respond to clock rate change

Buffer delay

When a statistical time division multiplexer is receiving more characters than it can send, it stores the excess in its random access memory. Normally, separate RAM buffers are used for transmission and reception of data, with each buffer shared by all channels. This design method results in two buffer pools of memory in each STDM, enabling active channels to use memory that would otherwise be idle, and is preferable and less costly to individual buffer memory assigned to each channel.

As buffer memory fills due to high activity on many channels or transmission errors on a high-speed link connecting STDMs, causing retransmissions which delay throughput through the multiplexer's buffer memory, buffer occupancy will rise and reach a predefined level. This level, referred to as high buffer occupancy, is about 80% of available buffer memory. When reached, this level results in the STDM initiating flow control. Some STDMs will initiate flow control on all active channels, while other STDMs prioritize flow control by first disabling the highest operating rate data sources.

Once flow control is enabled, the level of buffer occupancy will begin to decrease. When occupancy reaches a predefined low level, referred to as low buffer occupancy, flow control is disabled, enabling data sources attached to the STDM to resume transmission. Usually the low buffer occupancy is set at approximately 30% of buffer capacity.

When an STDM has an unusually high number of active channels, or transmission errors between STDMs occur frequently, the enabling and disabling of flow control will also occur frequently. During this time the level of buffer occupancy will toggle between approximately 30% and 80% of available buffer memory.

In addition to the enabling of flow control causing a transmit delay through an STDM, the occupancy level of the transmission buffer also affects throughput performance. This is because once flow control is disabled, servicing occurs on a first-in, first-out basis. Thus, if there were X characters in the transmission buffer, the next character sent into the STDM has to wait until X characters are output onto the high-speed line until that character flows onto the line.

We can approximate the transmission buffer delay by first obtaining the average buffer occupancy level through the use of a statistics port built into many STDMs. By multiplying the average buffer occupancy level in characters by 120% and dividing the result by the high-speed line operating rate we can obtain the average transit delay through the STDM. The reason we should multiply the occupancy level by 120% is based upon the assumption that the frame format results in approximately a 20% overhead.

As an example of computing buffer transit delays, let us assume that through the use of an STDM's statistics port we determined its average transmit buffer occupancy level to be

3000 characters. Let us further assume the STDM is conected to a modem operating at 19.2 kbps. Then, the average transmit buffer delay becomes:

$$\frac{3000 \text{ characters} \times 1.2}{19\,200 \text{ bps} \div 8 \text{ bits/character}} = 1.5 \text{ seconds.}$$

Although the transmit buffer capacity of most STDMs varies between a few thousand and 64K characters in size, it is not only the actual size of the buffer but its average occupancy level and the number of flow controls issued during a period of time that should be of concern. A large buffer enables a larger quantity of data to be buffered prior to the STDM enabling flow control; however, when flow control is implemented the higher level of buffer occupancy results in an increase in the transit delay time through the STDM. A smaller buffer results in a smaller quantity of data being buffered prior to the STDM enabling flow control. Although this reduces the transit delay through the STDM, when a large number of channels are active or the error rate on the circuit connecting STDMs is high, the enabling and disabling of flow control occurs more frequently. For the terminal user this results in random delays, where portions of screen activity suddenly stop and resume in correspondence to the enabling and disabling of flow control. Although the size of available buffer memory results in a trade-off between transit delay and flow control generation, some STDMs use buffer memory more efficiently than others. Such STDMs use a priority scheme for issuing flow control and servicing data sources on different channels. For example, the STDM would flow control a streaming channel to prevent it from locking out other channels and might segment the contents of buffer memory to enable buffered data from other channels to reach the multiplexer's frame prior to buffered data from the streaming channel.

4.3 SERVICE RATIOS

The increased efficiency of STDMs in comparison to traditional TDMs is normally expressed by a term known as the STDM's service ratio. Here, the service ratio is simply the multiplier of input data sources in bits per second into a statistical multiplexer in comparison to a traditional time division multiplexer.

Examples

One example of the utilization of service ratios can be obtained from considering the conventional TDM illustrated in the top of Figure 4.5. If the multiplexer is connected to a 9600 bps modem, its maximum aggregate input is also 9600 bps. Thus, a maximum of four 2400 bps data sources, or a combination of other data sources adding up to 9600 bps, defines the input servicing rate of this multiplexer.

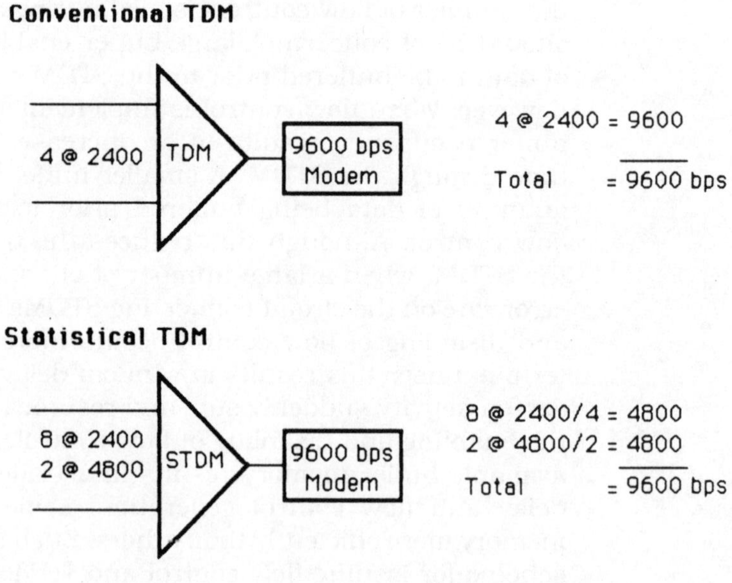

Conventional TDM

4 @ 2400
TDM
9600 bps Modem
4 @ 2400 = 9600
Total = 9600 bps

Statistical TDM

8 @ 2400
2 @ 4800
STDM
9600 bps Modem
8 @ 2400/4 = 4800
2 @ 4800/2 = 4800
Total = 9600 bps

Figure 4.5 Comparing statistical and conventional TDMs. A statistical multiplexer typically has an efficiency of 1.5 to 4 times that of a conventional TDM. Assuming a service ratio of 4:1 for asynchronous data and 2:1 for synchronous data, in this example the STDM can support twice the number of asynchronous data sources as the TDM as well as service two synchronous data sources.

The increased efficiency of statistical multiplexers based upon their ability to strip start, stop, and parity bits from asynchronous data, ignore idle channels, create variable-length channel data onto the multiplexer frame and, in some instances, perform data compression, permits STDMs to service additional data sources. This additional input servicing capacity is expressed by the device's service ratio. For asynchronous transmission, most STDMs have a service ratio between 2:1 and 4:1, permitting two to four times the aggregate input data rate of traditional TDMs. For synchronous transmission, most

STDMs have a service ratio between 1.5:1 and 2:1. This lower service ratio is due to the greater efficiency of synchronous transmission over asynchronous transmission, which normally cannot have bits stripped from each character and whose time interval between characters is zero when the data source is active. Thus, the lack of gaps between characters when synchronous data sources are active contributes to a lower service ratio.

The lower portion of Figure 4.5 illustrates an example of an STDM servicing asynchronous and synchronous data sources when connected to a 9600 bps modem. In this example, it was assumed that the STDM has a service ratio of 4:1 for asynchronous data and 2:1 for synchronous data. Based upon these service ratios, if the eight 2400 bps data sources represent asynchronous transmission, their effective input is $8 \times 2400/4$, or 4800 bps. Similarly, if the two 4800 bps data sources represent synchronous transmission, their effective input is $2 \times 4800/2$, or 4800 bps. Together, the effective asynchronous and synchronous data sources match the operating rate of the modem.

Data source support

Some statistical multiplexers only support asynchronous data while other multiplexers support both asynchronous and synchronous data sources. When a statistical multiplexer supports synchronous data sources it is extremely important to determine the method used by the STDM vendor to implement this support.

Some statistical multiplexer vendors employ a bandpass channel to support synchronous data sources. When this occurs, not only is the synchronous data not multiplexed statistically, but the data rate of the synchronous input limits the overall capability of the device to support asynchronous transmission.

Figure 4.6 illustrates the effect of multiplexing synchronous data via the use of a bandpass channel. When a bandpass channel is employed, a fixed portion of each message frame is reserved for the exclusive multiplexing of synchronous data, with the portion of the frame reserved proportional to the data rate of the synchronous input to the STDM. This means that only the remainder of the message frame is then available for the multiplexing of all other data sources.

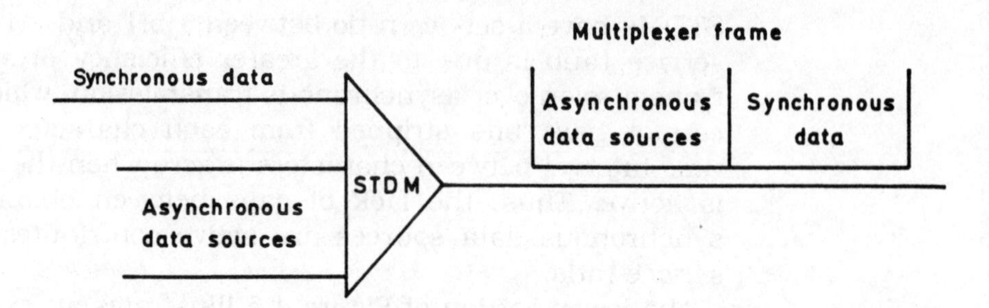

Figure 4.6 The use of a bandpass channel to multiplex synchronous data. The synchronous data source is always placed into a fixed location on the multiplexer frame. In comparison, all asynchronous data sources contend for the remainder of the multiplexer frame.

As an example of the limitations of bandpass multiplexing, consider an STDM that is connected to a 9600 bps modem and supports a synchronous terminal operating at 7200 bps. If bandpass multiplexing is employed, only 2400 bps is then available in the multiplexer for the multiplexing of other data sources. In comparison, assume another STDM statistically multiplexes synchronous data. If this STDM has a service ratio of 1.5:1, then a 7200 bps synchronous input to the STDM would on the average take up 4800 bps of the 9600 bps operating line. Since the synchronous data is statistically multiplexed, when that data source is not active other data sources serviced by the STDM will flow through the system more efficiently. In comparison, the bandpass channel always requires a predefined portion of the high-speed line to be reserved for synchronous data, regardless of the activity of the data source.

ITDMs

One advancement in statistical multiplexer technology has resulted in the introduction of data compression into a few STDMs. Such devices intelligently examine data for certain characteristics and are known as intelligent time division multiplexers (ITDM). These devices take advantage of the fact that different characters occur with different frequencies and use this quality to reduce the average number of bits per character by assigning short codes to frequently occurring characters and long codes to seldom-encountered characters.

The primary advantage of the intelligent multiplexer lies in its ability to make the most efficient use of a high-speed data circuit in comparison to the other classes of TDMs. Through compression, synchronous data traffic which normally contains minimal idle times during active transmission periods can be boosted in efficiency. Intelligent multiplexers typically permit an efficiency four times that of conventional TDMs for asynchronous data traffic and twice that of conventional TDMs for synchronous terminals.

4.4 STDM STATISTICS

Although the use of statistical multiplexers can be considered on a purely economic basis to determine if the cost of such devices is offset by the reduction in line and modem or DSU costs, the statistics that are computed and made available to the user of such devices should also be considered. Although many times intangible, these statistics may warrant consideration even though an economic benefit may at first be hard to visualize. Some of the statistics normally available on statistical multiplexers are listed in Table 4.2. Through a careful monitoring of these statistics, network expansion can be preplanned to cause a minimum amount of potential busy conditions to users. In addition, frequent error conditions can be noted prior to user complaints and remedial action taken earlier than normal when conventional devices are used.

Table 4.2 STDM statistics

Multiplexer loading: % of time device not idle
Buffer utilization: % of buffer storage in use
Number of frames transmitted
Number of negative acknowledgements (NAKs) received

$$\text{Error density} = \frac{\text{NAKs received}}{\text{frames transmitted}}$$

$$\text{Compression efficiency} = \frac{\text{total bits received}}{\text{total bits compressed}}$$

$$\text{Statistical loading} = \frac{\text{number of actual characters received}}{\text{maximum number which could be received}}$$

$$\text{Character error rate} = \frac{\text{characters with bad parity}}{\text{total characters received}}$$

Multiplexer loading

Multiplexer loading is normally supplied as a percentage of a time period during which the device was not idle. By itself this statistic may not be meaningful, as you would expect a high percentage of loading during the workday and a low percentage or zero percentage of loading on weekends and in the evenings.

Buffer utilization

Buffer utilization statistics normally show the percentage of occupancy of the local and remote transmit buffers. If you have a high percentage of buffer utilization, your STDMs are probably enabling and disabling flow control frequently. If you have a low percentage of buffer occupancy, it is reasonable to expect a low level of enabling and disabling of flow control, resulting in users obtaining a smoother data flow through the multiplexers. If your STDM does not provide flow control statistics you can use buffer utilization as a general indication of flow control activity.

Buffer occupancy

In place of providing a percentage of buffer utilization some STDMs provide the average transmit buffer occupancy in characters for a previously defined time frame. This time frame normally functions as a sliding window. That is, if it is 15 minutes, each access to the buffer utilization report provides the average transmit buffer occupancy for the past 15 minute interval. As previously discussed, you can use this information to compute the transit delay through the multiplexer to determine if excessive STDM loading is resulting in poor response time performance.

Frame transmitted and negative acknowledgements

The number of frames transmitted and number of negative acknowledgements (NAKs) received are usually provided as counts for a predefined period of time. By themselves they are not very meaningful, however, dividing NAKs received by frames transmitted provides a measurement known as error density which provides an indication of the quality of the line connecting two multiplexers.

If network users complain about excessive response times when accessing a computer through STDMs, the error density used in conjunction with flow control statistics may be used to isolate the cause of the problem. For example, if flow control levels are high and error density is low, the probable cause of excessive response times is an excessive transmission loading on the multiplexers. Conversely, if both the error density and flow control levels are low, this indicates that the problem may reside in the computer being accessed. Thus, the second situation would indicate that both communications equipment and the transmission facility are functioning at an acceptable level of performance and the resolution of the reported problem lies elsewhere.

Ratios

In addition to providing statistics, many STDMs also compute a number of ratios which provide valuable peformance measurement information. The four ratios listed at the bottom of Table 4.2 are but a few of the many ratios collectively reported by STDMs manufactured by different vendors with the ratios reported by a specific STDM dependent upon the vendor's inclusion of data recording elements which enable the ratio to be computed.

Compression efficiency

The compression efficiency ratio provides an indication of the susceptibility of data flowing through the STDMs to the compression method used by the multiplexers. This information can be valuable in determining what additional data sources, if any, can be connected to a STDM prior to bottlenecks occurring. For example, a compression ratio of 3 would indicate that the connection of a 2400 bps device would add 800 bps under worst case conditions when the new device is performing a file transfer operation or printing a large file transmitted through the STDM system.

Statistical loading

The statistical loading ratio is more valuable than the multiplexer loading percentage as it indicates whether or not users communicating through the multiplexer are being adversely delayed. In general, a loading above 75% will result in transmission delays since framing overhead will cause the aggregate of data transmitted to exceed the transmission capacity of the line, resulting in buffer occupancy increasing. If your STDM does not provide a direct measurement of buffer utilization you can use statistical loading as an indirect indication of buffer utilization.

Character error rate

The character error rate ratio is only applicable for asynchronous data sources. This ratio provides you with an indication of line quality from a terminal location accessing an STDM location to the STDM.

Many modern STDMs provide both statistics and ratios on an aggregate basis for the entire multiplexer as well as on an individual basis for each port when the statistic or ratio is applicable to an individual port. For example, an STDM might indicate flow controls issued to each port during a specified period of time as well as the total number of flow control signals issued during that period of time. By carefully analyzing the statistics and ratios provided by STDMs, you not only obtain the ability to achieve an insight into network performance, but, in addition, may observe potential problems and initiate corrective action prior to users experiencing those problems.

4.5 STDM FEATURES

The list of features either included with statistical multiplexers or offered as options varies considerably between different vendor products as well as within a vendor's product line. In Table 4.3, the reader will find a list of STDM features offered by many vendors. Since some of these features are also available on traditional TDMs, Table 4.3 also indicates the applicability of each feature to both types of multiplexers.

Auto speed and auto code detection

Auto speed detection permits the multiplexer to automatically adjust to the speed of asynchronous data sources. This feature

Table 4.3 Common multiplexer features

	Applicability	
Feature	STDM	TDM
Auto speed detection	yes	yes
Auto code detection	yes	yes
Audio alarm	yes	yes
Auto echo	yes	yes
Bandpass	yes	no
Command console operation	yes	no
Data compression	yes	no
Flyback delay	yes	yes
Protocol support	yes	yes
Port contention and switching	yes	no
Multi-node operation	yes	no
Redundant common logic	yes	yes
Redundant power supply	yes	yes
Test cards	yes	yes
Statistics display	yes	no

permits the multiplexer to dynamically configure itself and enables a single channel to support a variety of asynchronous devices without an operator having to manually reconfigure the multiplexer each time a different device transmits data into a channel. Similar to auto speed detect, auto code detect enables a multiplexer to dynamically adjust itself to service different data codes on a common channel.

Auto echo

Auto echo, which is also referred to as echoplex, permits a multiplexer to serially echo received data from an asynchronous device back to that device. Its utilization provides a primitive form of error control, since it enables the user to see characters transmitted to the multiplexer as they are echoed back to the terminal. In addition, by echoing the character back at the multiplexer channel, an echo at the computer port becomes unnecessary. Thus, the auto echo feature can also boost the performance of the multiplexer, since it eliminates a portion of the data traffic previously routed through a multiplexer system.

Bandpass

The bandpass feature on a statistical multiplexer removes the data rate used by the bandpass from statistical multiplexing. This feature enables multidrop lines and other time-critical data sources to be routed through the multiplexer with a minimum time delay.

Command console

The ability to connect a console to a multiplexer is normally associated with the capability to network multiplexers. Although this capability varies considerably between vendors, normally a networking capability enables an operator to dynamically assign the flow of data from one multiplexer through one or more intermediary devices to a specific channel on another multiplexer. This capability can be used for alternate routing if a line between multiplexers should fail or to establish paths to multiple computers that end-users can select by entering an appropriate code.

Data compression

The ability to compress data results in a reduction in the quantity of data which flows through an STDM system. This in turn enables STDMs to carry additional traffic or reduces the delays through STDMs in comparison to the use of statistical multiplexers that do not compress data.

Flyback delay

On electromechanical printers, a delay time is required between sending a carriage return to the terminal and then sending the first character of the next line to be printed. This delay time enables the print head of the terminal to be repositioned prior to the first character of the next line being printed. Many statistical multiplexers can be set to generate a series of fill characters after detecting a carriage return, enabling the print head of an electromechanical terminal to return to its proper position prior to receiving a character to be printed. This feature is called flyback delay and can be enabled or disabled by channel on many multiplexers.

Protocol support

The protocols supported by a multiplexer govern its capability to support terminal devices that transmit data according to those protocols. While IBM's bisynchronous and synchronous data link control (SDLC) protocols are supported by many vendors, some vendors also support a multitude of other protocols, including Honeywell 7700, Sperry UTS 200, Wang, and other vendor-specific transmission sequences.

Most STDM vendors segregate the support for different basic types of protocols to different types of channel adapters. For example, different types of bisynchronous protocols are normally supported by a bisynchronous adapter card while bit-oriented protocols are supported by the use of a bit-oriented adapter card. Since most channel adapter cards are manufactured to support multiple data sources and since all STDMs have a limit to the number of channel adapter cards they support, the mixture of protocols used by different data sources is a further constraint you may have to consider when using STDMs.

Port contention

Port contention is an option that enables an STDM to permit originating data sources to contend for one or more of a group of channels on a distant multiplexer. This option enables the multiplexer to obtain a portion of the capability of a stand-alone port selector in addition to its multiplexing role. Some multiplexers build upon this feature, permitting switching between channels, which permits the networking of multiplexers to obtain virtual routes that are assigned based upon end-user requests.

Multi-node

The multi-node feature permits a multiplexer to support two or more high-speed composite channels. This feature also enables multiplexers to perform alternate routing if a high-speed circuit should fail.

Statistics display

The statistics display permits users to monitor the performance of the multiplexer and to adjust its configuration when necessary. Typical statistics displayed upon demand can include the percentage of the time the device is not idle, the activity of individual channels in the multiplexer, and the utilization and error rate of the high-speed circuit or circuits the device interfaces.

The other features listed in Table 4.3 govern the ability of the multiplexer to set an alarm when a predefined failure occurs and support the installation of specific diagnostic test equipment and various types of redundant equipment. Since these features are self-explanatory, they are not discussed in this section.

4.6 STDM APPLICATIONS

The ability of frames to carry addressing information enables STDMs to expand upon the networking capability of TDMs. Thus, STDMs can perform all of the TDM networking functions described in Chapter 3 as well as many topological variations unique to STDMs. In this section we will examine a few of the more interesting topological configurations obtainable from the use of STDMs based upon their addressing capability as well as the ability of some STDMs to support multi-nodal connections to other STDMs.

Multi-node networking

One of the more interesting capabilities of STDMs is the ability of some multiplexers to support multiple data links. This capability enables other STDMs to be located where clusters of terminals exist as illustrated in Figure 4.7. In this example, we assumed eight terminal devices are located in both New York and Chicago and require access to a computer located in Atlanta. Instead of obtaining four multiplexers, we could install a multi-nodal STDM, referred to by some vendors as an intelligent network processor, in Atlanta. Doing so enables a single STDM in Atlanta to service both New York and Chicago STDM users.

Multi-nodal STDMs normally support up to eight data links through the use of individual link modules. Since a failure of a

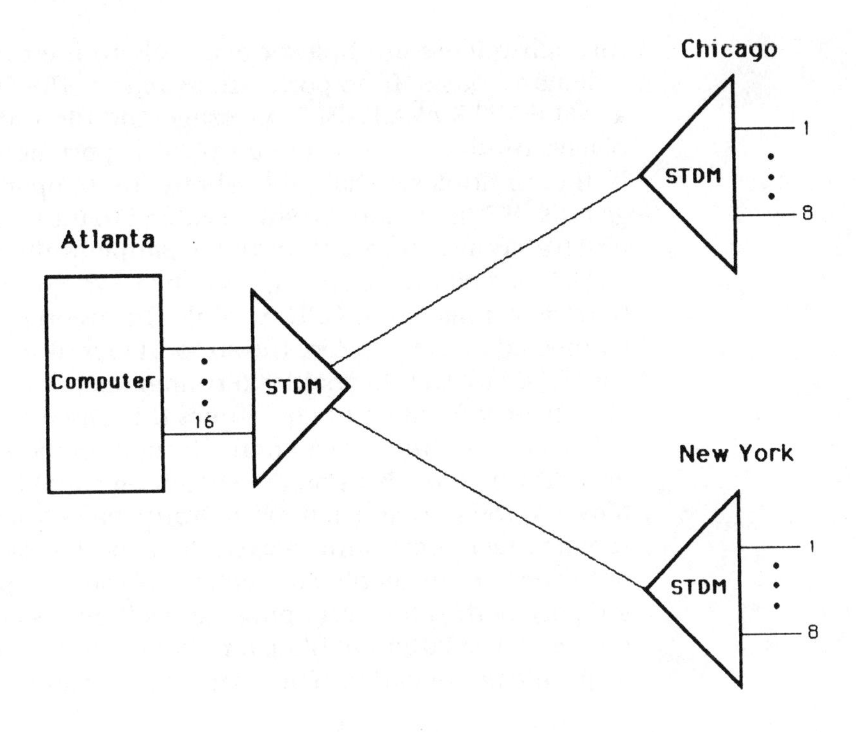

Figure 4.7 The use of a multi-nodal STDM enables two or more remotely located STDMs to share connectivity to a common multiplexer. In this example, the use of a multi-nodal STDM in Atlanta reduces the number of multiplexers required in this network from four to three.

multi-nodal STDM can render an organization's entire network or a significant portion of the network inoperative, most vendor's market dual common logic and redundant power supplies for this type of equipment. If the multi-nodal multiplexer has a port contention and switching capability the STDM can be used to provide access via a common network to multiple computers as well as enable complex networks to be developed.

Port contention is normally incorporated into large-capacity multi-nodal statistical multiplexers that are designed for installation at a central computer facility. This type of STDM may demultiplex data from hundreds of data channels, however, since many data channels are usually inactive at a given point in time, it is a waste of resources to provide a port at the central site for each data channel on the remote multiplexers. Thus, port contention results in the STDM at the central site containing a lesser number of ports than the number of channels of the distant multiplexers connected to that device. Then, the STDM at the central site contends the data sources entered through remote multiplexer channels to the available ports on a demand basis. If no ports are available, the STDM may issue a 'NO PORTS AVAILABLE' message and disconnect the user or places the user into a queue until a port becomes available. If the multiplexer has the ability to support multiple port groups, STDM channels can be cabled to multiple computers as illustrated in Figure 4.8. In this example we have assumed the STDM located in Atlanta supports two port groups. One group of ports is connected to CPU A, while the second group of ports is connected to CPU B. Here the support of multiple port groups by the STDM in Atlanta enables terminal users in Chicago and New York to access different computers through a common network.

The actual routing of a channel from a remote location onto a port group depends upon the switching capability of the STDM. Most STDMs treat each port group as a rotary, enabling a remote user requesting access to a port group to be placed onto the first available connection in the group. Other STDMs support both rotary groupings as well as specific port-routing requests, the latter enabling a remote user to request routing to a specific port location. The actual assignment of port groupings

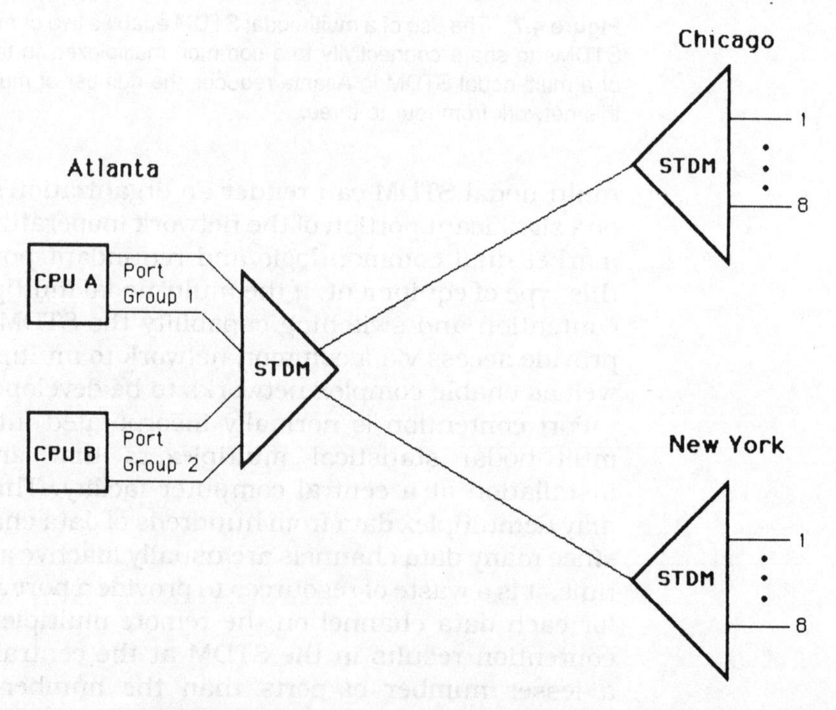

Figure 4.8 STDM port group support. By supporting port groupings, an STDM can be used to route incoming data traffic to multiple computer systems.

is performed at the switching STDM. To facilitate the use of this capability, most switching STDMs support the assignment of English-like commands to port groups. For example, HP3000 could be used to support the switching of a remote user to a port group connected to a Hewlett-Packard HP3000 computer, while IBM3081 could be used to provide access to a port group connected to an IBM 3081 computer.

Building STDM networks

Although the primary function of an STDM is to reduce communications costs by multiplexing data, this category of equipment may also be employed to obtain sophisticated networking capability. This additional capability is obtained by using several STDM features, such as multi-nodal support, port contention, and data switching together.

Figure 4.9 illustrates an example of the use of three STDMs to develop a complex network whereby users entering any STDM can request access to one of four computer systems connected to the network. In this example, each STDM network node is connected via a high-speed circuit to two other nodes, permitting alternate routing to occur in the event that any circuit or modem should fail. In such situations, data would be automatically routed on an alternate path through an intermediate STDM to its destination. At node 3, it was assumed that the STDM has a port contention feature. In this example, it was also assumed that the multiplexer's output is partitioned into two ports or routing groups. One group of ports is connected to one computer at that node, while the second group of ports is connected to a second computer at that node.

The STDM network illustrated in Figure 4.9 permits the routing of any input data source to one of four computers connected to the network. This routing capability is based upon the assumed switching and port contention capability of the STDMs. When these features are available in the multiplexers, one network user is required to be a system or network administrator. Using a command console connected to one STDM, this individual is responsible for setting up routing groups between multiplexers, as well as the alternate routes that are automatically established in the event that a circuit between multiplexers should become inoperative.

As indicated in this chapter, statistical multiplexers provide a significant level of performance over traditional TDMs. Thus,

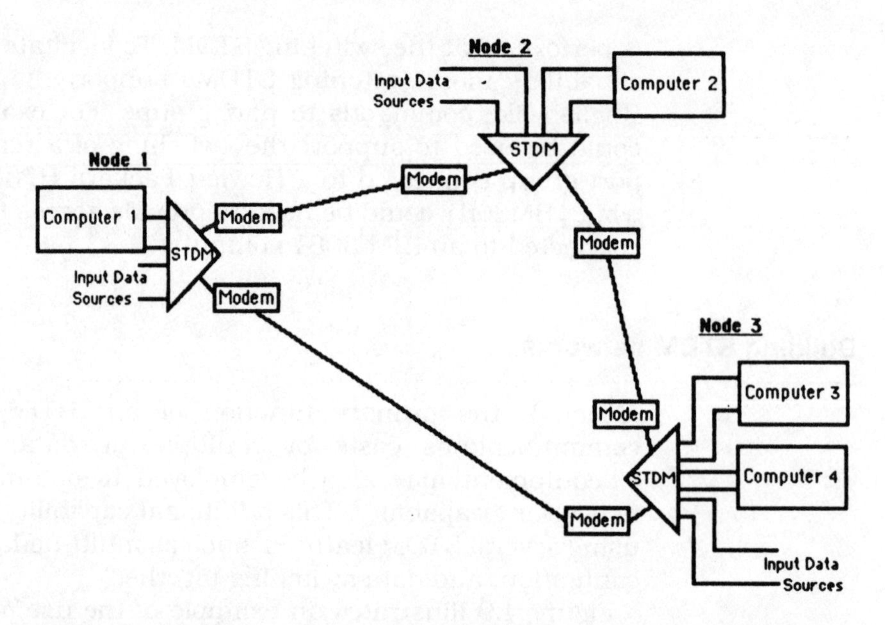

Figure 4.9 Networking STDMs. When STDMs have multi-nodal support and the ability to create virtual channels they can be used to network several computer sites together. If alternate routing capability is included in each multiplexer, the failure of one data link is compensated by the alternate routing of data sources through an intermediate multiplexer.

one question readers may have in the back of their mind is 'why aren't all TDMs manufactured as STDMs?' To answer this question consider some of the differences between TDMs and STDMs in the areas of transit delay, use of flow control, and variable throughput based upon traffic occurring on other channels.

A TDM reserves a slot for the specific use of each data source. While not efficient, this technique ensures the transit delay through the multiplexer is negligible. In addition, this technique alleviates the necessity of flow control and provides a consistent level of throughput. For many applications, including the servicing of multidrop lines, integration of voice and data, and support of video, delays may not be acceptable. In such situations, conventional TDMs are usually preferable to the use of STDMs and this explains the reason why most T-carrier multiplexers are based upon TDM technology. Thus, although STDMs represent a significant addition to multiplexing technology, we can expect traditional TDMs to continue to be used in many communications applications where buffering delays cannot be tolerated.

5

PACKET ASSEMBLER/ DISASSEMBLER

In this chapter we will focus our attention upon a special type of statistical multiplexer known as a packet assembler/disassembler, or PAD. In actuality, a PAD can be represented by a software module operating on any type of computer, ranging in scope from a microprocessor to a front-end processor or mainframe computer. In fact, the first types of PADs were based upon the use of minicomputers and were designed to provide access from non-packet mode devices, primarily asynchronous terminals, to a packet switching network. Later, software modules were developed to operate on front-end processors and mainframes which enabled those devices to be directly connected to packet networks.

5.1 EVOLUTION

In the late 1980s, the increased capacity and computational capability of microprocessors enabled some vendors to manufacture PADs on an adapter board that could be inserted into the system unit of a personal computer. This was shortly followed by a few modem vendors incorporating the functionality of a PAD into their products.

Originally, the development of the PAD was based upon a requirement to permit multiple non-packet mode terminals at one location to obtain access to a packet switching network via a common leased line. Thus, the first type of PAD to be marketed was a multiple port device. Later, vendors developed single board devices that enable one personal computer to access a packet network via a leased line or a dial-up connection to an

X.25 packet port on a packet network. In tandem with physical design changes in PAD development was an evolutionary change in software. Originally, PADs were designed to convert asynchronous start–stop protocols to an X.25 protocol. Later, several vendors expanded the functionality of PADs by adding additional protocol support to these devices.

5.2 CCITT X SERIES RECOMMENDATIONS

To understand the operation and functionality of a PAD requires a short review of several CCITT X Series recommendations which deal with packet switching networks, including the X.25, X.3, X.28, and X.29 recommendations.

X.25

The CCITT X.25 recommendation specifies the interface between data terminal equipment (DTEs) and data circuit-terminating equipment (DCEs) for DTEs operating in a packet mode on a public data network. In this context, DTEs can be any type of computer or terminal device, such as a front-end processor or personal computer capable of supporting the X.25 protocol. Here the DCE references the public data network nodes or switching equipment which provide access to the network. Figure 5.1 provides an overview of the CCITT X.25

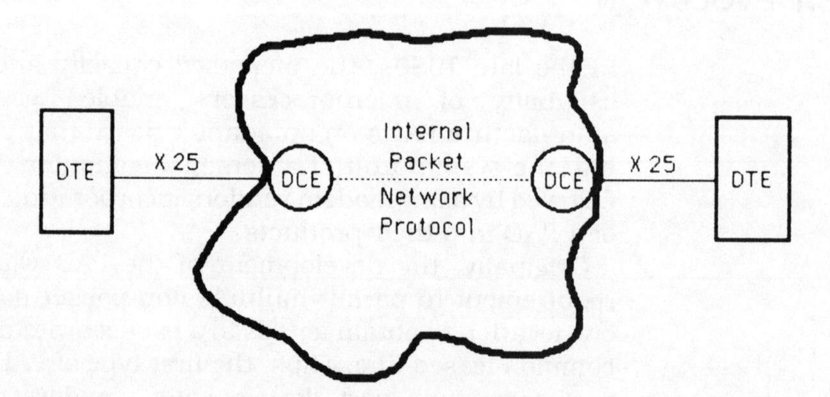

Figure 5.1 X.25 DTE and DCE relationship. Under the CCITT X.25 recommendation, a DTE is any device capable of supporting the X.25 protocol.

recommendation, illustrating the relationship between DTEs and DCEs under the context of this recommendation.

Under the X.25 recommendation there are three distinct levels or layers of control which correspond to the Open System Interconnection (OSI) reference model. These levels are the physical level (level 1), the data or link level (level 2), and the packet or network level (level 3).

Physical level

The physical level defines the interface required for the transmission of data on a synchronous full-duplex circuit connecting a DTE to DCE. The primary recommendation defined by the CCITT is X.21 which is an intelligent interface; however, recommendation X.21bis was promulgated as a secondary recommendation for an interim period and represents the well established RS-232/V.24 interface.

Data or link level

The data or link level defines the logical interface between the DTE and the DCE. This level consists of procedures for controlling the flow of information, including link setup, packet transfer, and link disconnect. In addition, this level controls the transmission of frames of information between the DTE and DCE in an error-free mode due to the use of a built-in error detection and correction scheme which employs cyclic redundancy checking.

Packet level

The packet level enables the concurrent operation of multiple user calls over a single physical circuit. Included in the formation of packets is a unique logical channel number (LCN), which identifies each call for the duration of the call.

When the X.25 recommendation was promulgated, by far the vast majority of DTEs were non-packet mode devices. To accommodate the support of the most popular type of non-packet mode DTEs, asynchronous terminals, the CCITT also promulgated recommendations X.3, X.28, and X.29.

X.3

The CCITT X.3 recommendation defines a set of 22 parameters which govern the operation and functionality of a PAD. These parameters are used to support the connection of asynchronous devices to a packet network. The PAD parameters, which are described later in this chapter, can be set up and modified by either a remote computer or by the asynchronous device being serviced by the PAD.

X.28

The X.28 recommendation defines the procedures by which an asynchronous device interacts with the PAD. These procedures include the establishment of a physical connection between the device and the PAD, the initialization of service by the PAD, and the altering of PAD parameters, as well as the exchange of control information and user data between the PAD and an asynchronous device.

X.29

The third PAD-related recommendation is the CCITT X.29 recommendation. X.29 defines the procedures for the exchange of control information and user data between a local PAD and a packet mode DTE or between a local PAD and a remote PAD. Here the terms "local" and "remote" reference the location of the PAD with respect to a non-packet mode terminal device, with the local PAD providing a direct connection between that device and the packet network. The X.29 recommendation defines the procedures, including the formats of X.29 PAD messages and their function.

Figure 5.2 illustrates the relationship between the three previously described CCITT recommendations. In examining Figure 5.2, readers should note that the local PAD represents one location where this type of communications device can be used. In addition to its co-location with an asynchronous terminal, the PAD can reside in the device as an adapter board, in a modem connected to a terminal device, or can be located at the packet network node. Concerning the latter location, the PAD function is normally built into the packet network node and supports both non-packet mode terminal access via the

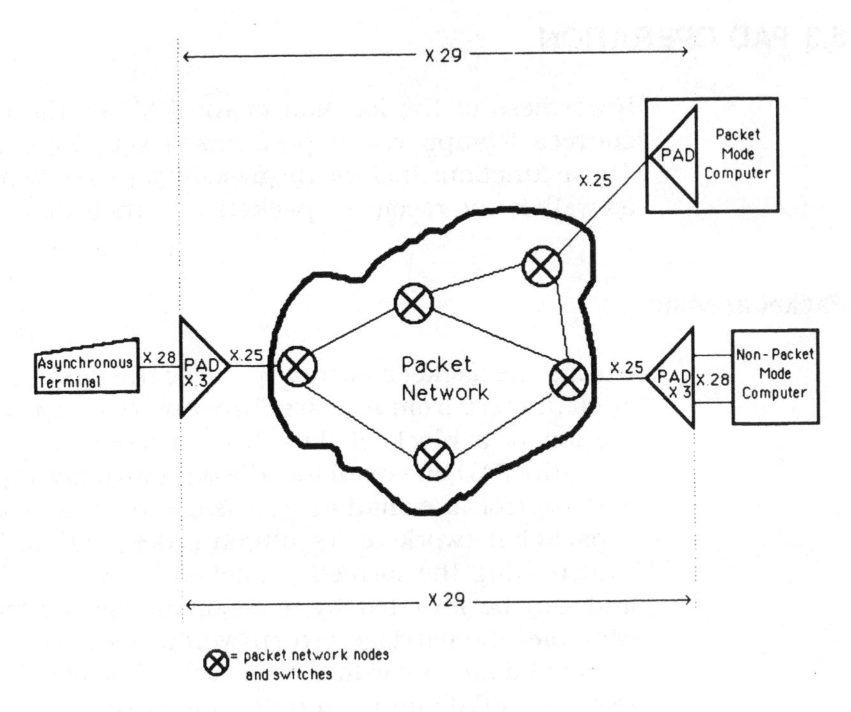

Figure 5.2 CCITT PAD-related recommendations.

switched telephone network as well as access via analog or digital leased lines. Since a packet node performs switching similar to a backbone switch in the network, for simplicity of illustration both nodes and switches were considered as being equivalent in Figure 5.2.

In examining Figure 5.2, note that two different types of computers can be connected to a packet network—an X.25 compatible computer and a non-X.25 compatible computer. If a terminal user accesses an X.25 compatible computer via a local PAD, the computer functions as a remote PAD. Then, control messages are exchanged between the local PAD and the X.25 compatible computer in the form of X.29 PAD messages as shown in the upper portion of Figure 5.2. If the computer being accessed is a non-X.25 device, it does not support X.25 nor X.29. Thus, a PAD must be used as an interface between the computer and the packet network. This is illustrated in the lower portion of Figure 5.2. Here the PAD is sometimes referred to as a host (computer) PAD. The host PAD operates similar to the local PAD servicing asynchronous terminals, and X.29 messages are then exchanged between the local PAD and the host PAD.

5.3 PAD OPERATION

Regardless of the location of the PAD or the number of data sources it supports, it performs a set of common functions. Those functions include the assembly and forwarding of packets as well as the receipt of packets and their disassembly.

Packet assembly

During the packet assembly process the PAD groups a collection of characters from an asynchronous data source or a block or portion of a block of data from a synchronous data source. Next, the PAD encapsulates the data with network identification and control information necessary to route a packet through a packet network to its ultimate destination. The process of transmitting the formed packet is known as data forwarding and can be initiated by a terminal user or by the PAD. For example, the carriage return would result in a terminal-user-initiated data forwarding, while the filling of a PAD buffer would result in a PAD-initiated buffer forwarding action.

When a data forward condition occurs, the PAD adds a packet header to the user data. The packet header contains three 8-bit bytes, referred to as octets, which includes identification, control, and sequencing information. Identification information includes a packet identifier number which is better known as the logical channel number (LCN). The LCN identifies each active data source being serviced by a multiple port PAD. Thus, we can say that a unique LCN represents each call.

Control information identifies the type of packet and, if applicable, specifies special network functions. This information is used by the packet network to determine the processing requirements for each packet.

The packet sequence number identifies the order in which packets are transmitted on each logical channel. This information is required for both packet acknowledgements as well as to provide information to the distant end of the correct disassembly of packets.

Packet framing

Once a packet is formed, it must be prepared for transport. This is accomplished by encapsulating the packet into an HDLC frame which wraps a header and trailer around the packet. The header contains link addressing as well as control and sequencing information which identifies the type of frame and specifies the data link function. The trailer is a two-octet CRC which serves as a mechanism for error detection. Figure 5.3 illustrates the relationship between user data flowing into the PAD, its packet formation, and encapsulation into a frame for transmission over an X.25 link.

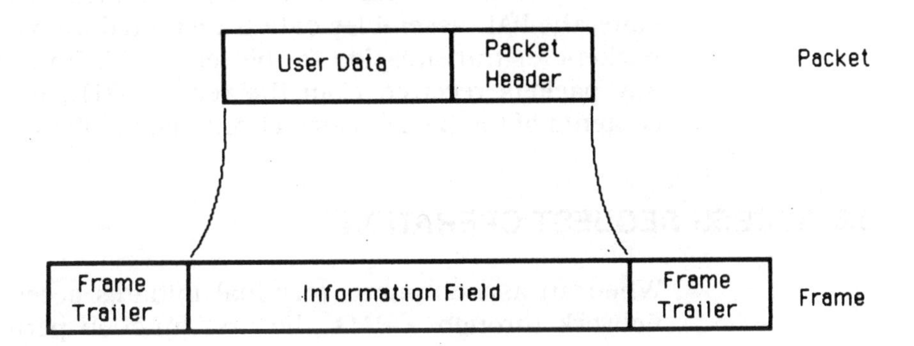

Figure 5.3 User data, packet, and frame relationship. User data is first packetized by a PAD. Next, the resulting packet is encapsulated into a frame for transmission.

5.4 PAD STATES

When a physical connection occurs between an asynchronous data source and a PAD, the PAD enters its command state. In this state, the terminal user transmits predefined characters or strings to the PAD which the latter recognizes as commands. Both PAD commands and reactions to those commands are defined by the CCITT X.28 recommendation.

Command state

During the time a PAD is in its command state, communications are limited to the flow of data between the terminal user and the PAD and the PAD and a remote DTE. Thus, the terminal user cannot directly communicate with the DTE when the PAD is in its command state.

During the time the PAD is in its command state, it responds to terminal user commands to establish a virtual call, clear a previously established call, or modify PAD parameters defined by the CCITT X.3 recommendation. During this time data transfer cannot occur between the terminal device and the DTE.

Data transfer state

Once a virtual call is established linking the terminal device to a distant DTE, the PAD will enter its data transfer state. In this state, the PAD assembles data transmitted from a terminal into packets for transmission to the remote DTE and disassembles any packets received from the remote DTE, transmitting the contents of the packets asynchronously to the terminal.

5.5 ACCESS REQUEST OPERATION

When an asynchronous terminal initiates access to a packet network through a PAD, the first function performed by the terminal user is the pressing of the carriage return key. Pressing this key results in the terminal transmitting a bit sequence to the PAD, which enables the PAD to identify the data code, speed, and parity setting of the terminal. Once this is accomplished, the terminal user and the PAD will exchange information. Typically, the PAD will transmit a herald message to the user, informing the user that they are connected to the network serviced by the PAD. This message is normally followed by a pair of numerics, typically separated by a dash, space, or colon, which identify the node and the port on the node to which the terminal user is connected. For example, 9043:12 might indicate node 9043, port 12.

After the node and port identifier are transmitted from the PAD to the terminal, the interaction between the terminal user and the PAD will be dependent upon the packet network serviced by the PAD. For example, consider the interaction between a terminal user and PADs connected to the BT Tymnet and US Sprint packet networks, illustrated in Figures 5.4 and 5.5, respectively.

BT Tymnet

Figure 5.4 illustrates the interaction of a Tymnet PAD with an asynchronous terminal user. Upon connection to the PAD, the

```
x~x'xxx'xxpx'px'xxxxxx xxpx^x^xx'xpx^x<'xpx'xppx'xxpx'xx^x
please type your terminal identifier
-6130:01-012-
please log in: cps


host:  WELCOME TO COMPUSERVE  1133


User ID:
```

Figure 5.4 Tymnet PAD access request operation.

```
TELENET
912 10B

TERMINAL=D1

@c 912140,opmoim,


912 140D CONNECTED
```

Figure 5.5 SprintNet (Telenet) PAD access request operation.

terminal user receives what appears to be a string of garbled characters. In actuality, that string is the message 'please type your terminal identifier' which is automatically output by the PAD at 300 bps and appears garbled when the terminal operates at a different speed.

When the terminal is set to operate at a data rate above 300 bps, the message appears garbled; however, entering two carriage returns at the end of the line enables the PAD to adjust the port to the terminal's operating characteristics. Thus, the second transmission of the previously mentioned message is readable.

The terminal identifier used in accessing BT Tymnet informs the PAD whether or not a delay is required after the carriage return character. Older electromechanical terminals require time for the print head to return to the leftmost position after a carriage return. Hence, the terminal identifier informs the PAD whether or not to insert null characters after each return. Once the terminal identifier is entered, the PAD returns the node and port number. In this example, the node (6130) has a node extension (01) and the port (012) represents the twelfth port on the node extension. This message is followed by the BT Tymnet 'please log in' message, which in effect asks the terminal user to enter the address for establishing a virtual call. In this example, the entry of the address CPS resulted in the establishment of a connection to CompuServe.

SprintNet (Telenet)

Figure 5.5 illustrates the interaction between a terminal user and a PAD when the PAD is connected to the US Sprint Sprint-Net (Telenet) packet network. In this example, the PAD responds to two carriage returns with the identifier TELENET, followed by the node (912) and port (10B) address. After the terminal user responds to the terminal identifier request with D1, the PAD generates an 'at' sign (@). This is a signal that the PAD is waiting for the entry of a user command. In this example, the user enters the c(onnect) command, with the letter c followed by the destination location (912140), identification (OPMOIM), and the password which is non-printable since the PAD is preconfigured not to echo passwords back to terminal users. Once that routing data is entered and a virtual call is established to the destination DTE, the message CONNECTED is returned by the PAD.

5.6 X.3 PARAMETERS

The key to the functionality of a PAD and its ability to support asynchronous data sources is the CCITT X.3 recommendation. This recommendation specifies the meaning, allowable values, and resulting PAD functionality for 22 PAD parameters. Table 5.1 lists the meaning of each of the PAD parameters defined by the CCITT X.3 recommendation.

The X.3 parameters listed in Table 5.1 should be considered as a nucleus of parameters upon which many vendors build additions to tailor their PADs to support required but currently undefined functions. For example, US Sprint SprintNet (Telenet) defines a parameter 24 referred to as 'Permanent Terminal.' This parameter can be assigned a value of 0 to have the PAD request the user's terminal identification or a value of 1 for the PAD to use a preset terminal identification. This

Table 5.1 CCITT X.3 PAD parameters

X.3 parameter number	X.3 meaning
1	PAD recall
2	Echo
3	Selection of data forwarding signal
4	Selection of idle timer delay
5	Ancillary device control
6	Control of PAD service signals
7	Procedure on receipt of break signal
8	Discard output
9	Padding after carriage return
10	Line folding
11	Binary speed
12	Flow control of the PAD by the DTE
13	Line feed insertion after carriage return
14	Padding after line feed
15	Editing
16	Character delete
17	Line delete
18	Line display
19	Editing PAD service signals
20	Echo mask
21	Parity treatment
22	Page wait

parameter enables SprintNet (Telenet) to enhance support for terminals directly connected to a PAD port by using a preset identifier. Otherwise, terminal users would be prompted for their terminal identification each time they require access to the packet network even though they are permanently connected to a port.

Each of the X.3 PAD parameters has two or more possible values. Initially, most PADs have their X.3 parameters established by the packet network's control center during the PAD installation process. Once the PAD's parameters are initialized, they may be changed through the use of a terminal connected to a command port of the device or, if allowed, by local and remote users connected to the PAD. In the remainder of this section, let us examine each of the X.3 defined parameters and their possible value.

1:m PAR recall

The setting of this parameter determines whether or not a terminal can initiate an escape from the data transfer mode to the command mode, enabling a command to be sent to the PAD after data transfer commences. Possible values are: 0 – escape not possible; 1 – escape possible.

2:m Echo

This parameter controls the echoing of characters on the user's screen as well as their forwarding to the remote DTE. Possible values are: 0 – no echo; 1 – echo.

3:m Selection of data forwarding signal

This parameter defines a set of characters that act as data forwarding signals when they are entered by the user. Coding of this parameter can be a single function or the sum of any combination of the functions listed below. As an example, a 126-code represents the functions 2 through 64, which results in any character, including control characters, being forwarded. Possible values of parameter 3 are: 0 – no data forwarding character; 1 – alphanumeric characters; 2 – character CR; 4 – characters ESC, BEL, ENQ, ACK; 8 – characters DEL, CAN, DC2; 16 – characters ETX, EOT; 32 – characters HT, LF, VT, FF; 64 – all other characters, X'00' to X'1F'.

4:m Selection of idle timer delay

This parameter is used to specify the value of an idle timer used for data forwarding. Possible values of this parameter include: 0 – no data forwarding on time-out; 1 – units of 1/20 second, maximum 255.

5:m Ancillary device control

This parameter enables flow control between the PAD and the terminal. The PAD uses the XON and XOFF characters (decimal 17 and 19) to indicate to the terminal whether or not it is ready to accept characters. Possible values are: 0 – no use of XON/XOFF (default); 1 – use XON/XOFF.

6:m Control of PAD service signals

This parameter provides for the suppression of all messages sent by the PAD to the terminal. Possible values of this parameter are: 0 – signals not transmitted (messages not sent); 1 – signals transmitted (messages sent) (default).

7:m Procedure on receipt of break signal

This parameter specifies the operation to be performed upon entry of a break character. Possible values of this parameter include: 0 – nothing; 1 – send an interrupt; 2 – reset; 4 – send an indication of break PAD; 8 – escape from data transfer state; 16 – discard output. Similar to parameter 3, parameter 7 can be coded as a single function or as the sum of a combination of functions.

8:m Discard output

This parameter controls transmission of data to the terminal. Possible values of this parameter are: 0 – normal data delivery to the terminal (default); 1 – discard all output to the terminal.

9:m Padding after carriage return

This parameter provides for automatic insertion of the PAD of null character padding after the transmission of a carriage return to the terminal. Possible values of this parameter are: 0 – no padding; 1–31 – 1–31 character delays.

10:m Line folding

This parameter provides for automatic insertion by the PAD of appropriate format effectors to prevent overprinting at the end of a terminal print line. Possible values of this parameter are: 0 – no line folding; *n* – characters per line before folding, where $1 \leq n \leq 255$.

11:m Binary speed

This parameter is set by the PAD when the terminal establishes a physical connection to the network. This allows the remote DTE or terminal user to examine the speed, as determined by the PAD. Decimal values are indicated in the following table.

Decimal value	Speed (bps)	Decimal value	Speed (bps)
0	110	10	50
1	134.5	11	75/1200
2	300	12	2400
3	1200	13	4800
4	600	14	9600
5	75	15	19 200
6	150	16	48 000
7	1800	17	56 000
8	200	18	64 000
9	100		

12:m Flow control of the terminal PAD

This parameter permits flow control of received data using XON and XOFF characters. Possible values are: 0 – no flow control; 1 – flow control.

13:m Line feed insertion after carriage return

This parameter instructs the PAD to insert a line feed (LF) into the data stream following each carriage return (CR). Possible values are: 0 – no LF insertion; 1 – insert an LF after each CR in the received data stream; 2 – insert an LF after each CR in the transmitted data stream; 4 – insert an LF after each CR in the echo to the screen. The coding of this parameter can be as a single function or a combination of functions by summing the values of the desired options.

14:m Padding after line feed

This parameter provides for the automatic insertion of padding characters into the character stream transmitted by the PAD to the start–stop DTE after the occurrence of a line feed character. This allows for the printing mechanism of the start–stop mode DTE to perform the line feed operation correctly and is only applicable in the data transfer state. Possible parameter values are: 0 – no padding after line feed; 1–255 – number characters inserted after line feed.

15:m Editing

This parameter permits the user to edit data locally or at the host. If local editing is enabled, the user can correct any data buffered locally; otherwise it must flow to the host for later correction. Possible values of this parameter include: 0 – no editing in the data transfer state; 1 – editing in the data transfer state.

16:m Character delete

This parameter permits the user to specify which character in the ASCII (International Alphabet Number 5) character set will be used to indicate that the previously typed character should be deleted from the buffer. Possible values for this parameter include: 0 – no character delete; 1–127 – character-delete character.

17:m Line delete

This character is used to enable the user to specify which character in the character set denotes that the previously entered line should be deleted. Possible values are: 0 – no line deleted; 1–127 – line-delete character.

18:m Line display

This parameter enables the user to define the character which will cause a previously typed line to be redisplayed. Possible values are: 0 – no line display; 1–127 – line-display character.

20:m Echo mask

This parameter is only applicable when parameter 2 is set to 1. When this occurs parameter 20 permits the user to specify which characters will be echoed. Possible values include: 0 – no echo mask (all characters echoed); 1 – no echo of character CR; 2 – no echo of character LF; 4 – no echo of characters VT, HT, FF; 8 – no echo of characters BEL, BS; 16 – no echo of characters ESC, ENQ; 32 – no echo of characters ACK, NAK, STX, SOA, EOT, ETB, and ETX; 64 – no echo of characters defined by parameters 16, 17, and 18; 128 – no echo of all characters in columns 0 and 1 of International Alphabet Number 5 and the character DEL.

21:m Parity treatment

This parameter controls how the PAD treats parity. Possible values are: 1 – parity checking; 2 – parity generation and 3 – parity checking and parity generation.

22:m Page wait

This parameter enables the PAD to suspend transmission of additional characters to an asynchronous DTE after a specified number of line feed characters have been transmitted by the PAD. Possible values of this parameter are: 0 – page wait disabled; 1–255 – number of line feed characters considered by the PAD for the page wait function.

5.7 X.28 COMMANDS

The key to the ability of a terminal user to display current X.3 parameters and, if desired, modify those parameters is the X.28 recommendation. This recommendation includes command signals sent by a terminal and recognized and acted upon by the PAD and service signals returned by the PAD to indicate predefined conditions have occurred.

Table 5.2 lists the format and function of X.28 PAD command signals. To illustrate the use of these command signals consider the use of the PAR? command illustrated in Figure 5.6. In this example, after accessing a US Sprint SprintNet (Telenet) PAD and entering your terminal identifier you use the PAR? command to determine the value of the PAD's X.3 parameters. This results in the Telenet PAD returning the value of those X.3 parameters supported by Telenet. In addition to many PADs

Table 5.2 PAD command signals.

Command signal format	Function
CLR	To clear down a virtual call.
INT	To transmit an interrupt packet.
PAR? (list of parameters)	To request the current values of specified PAD parameters.
PROF (identifier)	To give to PAD parameters a standard set of values.
RESET	To reset the virtual call.
SET (list of parameters and values)	To set or change PAD parameter values.
SET? (list of parameters and values)	To request changing or setting of the current values of specified PAD parameters and to request current values of specified parameters.

```
TELENET
912 10E

TERMINAL=d1

@par?
PAR1:1,2:1,3:2,4:80,5:0,6:1,7:0,8:0,9:0,10:80,11:3,12:0,13:0,14:0,15:0,16:127,17
:24,18:18

@
```

Figure 5.6 Using the PAR command.

not supporting all X.3 parameters, readers should note that many PADs do not support all PAD command signals.

As an alternative to the previous example you can use the PAR? command to request the value of one or a group of parameters. For example, PAR?3 might return the response PAR3:2, which would indicate that the third X.3 parameter is currently set to a value of 2.

Two X.28 commands used for parameter modification are SET and SET?. The SET command is used to modify the value of defined X.3 parameters. For example, SET 1:1, 2:1 would set parameters 1 and 2 to a value of 1.

The SET? command can be viewed as a combination of the SET and PAR? commands. That is, this command causes the PAD to first modify its X.3 parameters based upon the parameters and parameter values entered in the SET? command. Next, the PAD transmits back to the user the parameters that were set and their new settings. Thus, through the use of the PAR, SET, and SET? commands you can display and modify X.3 parameters. This in turn provides you with the ability to tailor the functionality of a PAD port to your specific requirement.

6

T-CARRIER MULTIPLEXERS

In this chapter we will examine the operation and utilization of multiplexers designed to interface the North American T1 circuit which operates at 1.544 Mbps and the European E1 circuit which operates at 2.048 Mbps. Collectively, we will refer to this class of multiplexing equipment as T-carrier multiplexers. However, when we refer to a specific function related to a multiplexer used on a specific type of transmission facility we will reference the device as a T1 or E1 multiplexer.

Since an understanding of the operation and utilization of T-carrier multiplexers requires knowledge about the T1 and E1 circuits, we will first examine North American and European T-carrier transmission facilities. Using this information as a base will enable us to then focus our attention upon the operation and utilization of T-carrier multiplexers.

6.1 T1 CIRCUIT EVOLUTION AND OPERATION

The origination of T1 multiplexing dates to the early 1960s, when AT&T was developing a method to reduce the number of voice trunks routed between telephone company central offices. Instead of transmitting voice in its analog format, AT&T first digitized individual voice conversations and then multiplexed the resulting digital data onto a high-speed transmission facility.

The device used by the telephone company to digitize and multiplex analog voice signals is known as a digital channel bank, or D bank. The resulting multiplexed signal is a high-speed digital stream of binary ones and zeros that operates at

1.544 Mbps in the United States and is known as a T1 carrier. In Europe, a similar D bank operates at 2.048 Mbps on the European equivalent of the T1 carrier, which is referred to as an E1 carrier.

T1 was originally developed by AT&T for local communications within the telephone company network. It was designed to improve service to its customers while reducing transmission costs to the company. In 1974, T1 circuits were made available for non-telephone company use. At that time, AT&T tariffed T1 service for use by the Department of Defense. In 1977, AT&T retariffed T1 service, making it available to commercial organizations.

The first commercial T1 tariff was very expensive, essentially precluding its use by a majority of commercial organizations that could not utilize the full capacity of the transmission facility. It was not until early 1983 that T1 service began to be widely used. In that year, AT&T retariffed the service and renamed it ACCUNET T1.5. Since then, the use of T1 circuits has rapidly increased. This increase in T1 usage is due to the more economical cost of the service, as well as advances in T1 multiplexer technology that have enabled organizations to integrate voice, data, and video transmission over a common circuit at a cost considerably lower than that of operating separate facilities.

Channel banks

Channel banks used by telephone companies were originally analog devices. They were designed to provide the first step required in the handling of telephone calls that originated in one central office, but whose termination point was a different central office. The analog channel bank included frequency division multiplexing equipment, permitting it to multiplex, by frequency, a group of voice channels routed to a common intermediate or final destination over a common circuit.

The development of pulse code modulation resulted in analog channel banks becoming unsuitable for use with digitized voice. AT&T then developed the D-type channel bank which actually performs several functions in addition to the time division multiplexing of digital data. The first digital channel bank, known as D1, contained three key elements as illustrated in Figure 6.1. The codec, an abbreviation for coder–decoder, converted analog voice into a 64 kbps PCM encoded digital data

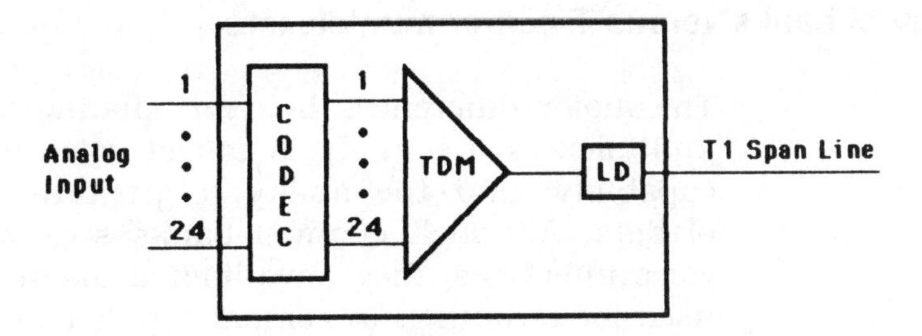

Figure 6.1 The D1 channel bank: TDM = time division multiplexer, LD = line driver.

stream. The TDM multiplexes 24 PCM encoded voice channels and inserts framing information to permit the TDM in a distant channel bank to be able to synchronize itself to the resulting multiplexed data stream that is transmitted on the T1 span line. The line driver conditions the transmitted bit stream to the electrical characteristics of the T1 span line, ensuring that the pulse width, pulse height and pulse voltages are correct. In addition, the line driver converts the unipolar digital signal transmitted by the multiplexer into a bipolar signal suitable for transmission on the T1 span line. Due to the operation of the digital channel bank, this equipment can be viewed as a bridge from the analog world to the digital world.

To ensure the quality of the resulting multiplexed digital signal, AT&T installed repeaters at intervals of 6000 ft on span lines constructed between central offices. Although repeaters are still required on local loops to a subscriber's premises and on copper wire span lines, the introduction of digital radio and fiber optic transmission has added significant flexibility to the construction and routing of T-carrier facilities.

Improvements made to the encoding method used by the D1 channel bank resulted in the development and installation of the D1D and D2 channel banks. Channel banks currently used include D3, D4, and D5, whose simultaneous voice channel carrying capacity varies from 24 to 96 channels.

Today, T1 lines are available from a variety of communications carriers, including AT&T, MCI, US Sprint, and others. In Europe, the equivalent T1 carrier, which is known as E1 and CEPT PCM-30, is available in most countries under different names. As an example, in the United Kingdom E1 service is marketed under the name MegaStream.

Channel banks versus T-carrier multiplexers

The major differences between channel banks and T-carrier multiplexers are in the areas of voice interfaces, diagnostic capability; and the ability to perform automatic rerouting of data. Although channel banks support a large variety of voice interfaces, they have limited diagnostics and cannot be used for rerouting. In comparison, T-carrier multiplexers may have a limited voice interface capability; however, they usually have superior diagnostics and many provide the capability to automatically reroute data.

Framing structure overview

In North American, the T1 carrier was designed to support the transmission of 24 channels of digitized voice.

Each channel is sampled 8000 times per second and 8 bits are used to represent the encoded height of the sampled analog wave. For synchronization, as well as other functions that are discussed later in this chapter, one framing bit is added to the digitized multiplexed data that represents 24 PCM encoded voice conversations.

Thus,

$$8 \text{ bits} \times 24 \text{ channels} + 1 \text{ framing bit} = 193 \text{ bits/frame}.$$

Since 8000 frames are transmitted each second, the bit rate is

$$193 \text{ bits/frame} \times 8000 \text{ frames/second} = 1.544 \text{ Mbps}$$

which is the operating rate of the North American T1 carrier facility.

In Europe, the T-carrier is commonly referred to as an E1 facility or a CEPT PCM-30 system, where CEPT is an acronym for the Conference of European Postal and Telecommunications, a European standards organization. CEPT uses a 32-channel system, where 30 channels are used to transmit digitized speech received from incoming telephone lines, while the remaining two channels are used for signal and synchronization information. Each channel is assigned a time slot as listed in Table 6.1.

The frame composition of an E1 or CEPT system consists of 32 channels of 8 bits per channel, or 256 bits per frame. No framing information is required to be added to the frame as in

Table 6.1 CEPT time slot assignments

Time slot	Type of information
0	Synchronization (framing)
1–15	Speech
16	Signaling
17–31	Speech

the North American T1 carrier since synchronization is carried separately in time slot zero.

Since 8000 frames per second are transmitted, the bit rate of an E1 facility becomes

$$256 \text{ bits/frame} \times 8000 \text{ frames/second} = 2.048 \text{ Mbps}.$$

Similar to the hierarchy illustrated for FDM in Figure 2.1, communications carriers have developed a hierarchy of digital carrier levels. Table 6.2 lists the digital carrier hierarchy levels in North America, Europe, and Japan.

In North America, T1C and T2 were developed to boost the carrying capacity of copper wire pairs beyond that obtained by using T1. These facilities are primarily restricted to use by telephone companies. T3, which operates at 44.736 Mbps, initially was offered to commercial organizations during 1988 and can be expected to grow in use.

The North American line types listed at the top of Table 6.2 actually reference the type of signal each line is capable of carrying. A T1 line carries a DS1 signal. Here DS1 (digital signal, level 1) is the 1.544 Mbps signal defined by AT&T to include pulse height and width, impedance, and other parameters. Although column two labeled 'Line signal standard' commences with DS1, in actuality, the lowest signal level in the digital hierarchy is DS0 (digital signal, level 0). DS0 refers to each 64 kbps digitized PCM data stream generated in a D-type channel bank, with 24 such DS0 channels along with framing bits used to form a DS1 signal.

Figure 6.2 illustrates the North American digital signal hierarchy. Note that DS0 originates at the digital channel bank located in the lower left portion of the illustration. The five data rates shown entering the data multiplexer in the upper left portion are Dataphone Digital Services transmission facilities, which are also commonly referred to as subrate services as

Table 6.2 Digital hierarchy levels

North America

Line type	Line signal standard	Number of voice circuits	Bit rate (Mbps)
T1	DS1	24	1.544
T1C	DS1C	48	3.152
T2	DS2	96	6.312
T3	DS3	672	44.736
T4	DS4	4032	274.176
T5	DS5		
T6	DS6		

Europe

Level number	System	Number of voice circuits	Bit rate (Mbps)
1	M1	30	2.048
2	M2	120	8.448
3	M3	480	34.368
4	M4	1920	139.264
5	M5	7680	565.148

Japan

Level number	System	Number of voice circuits	Bit rate (Mbps)
1	F-1	24	1.544
2	F-6M	96	6.312
3	F-32M	480	34.064
4	F-100M	1440	97.728
5	F-400M	5760	397.20
6	6F-4.6G	23040	1588.80

they operate below the DS1 rate. Dataphone Digital Services was originally developed by AT&T as an all-digital transmission facility for the transmission of data as opposed to voice. Since its introduction by AT&T, other communications carriers have introduced equivalent digital facilities for the transmission of data.

In the North American digital signal hierarchy, the M12 multiplexing system used by AT&T accepts four DS1 input signals and produces a DS2 output representing 96 DS0

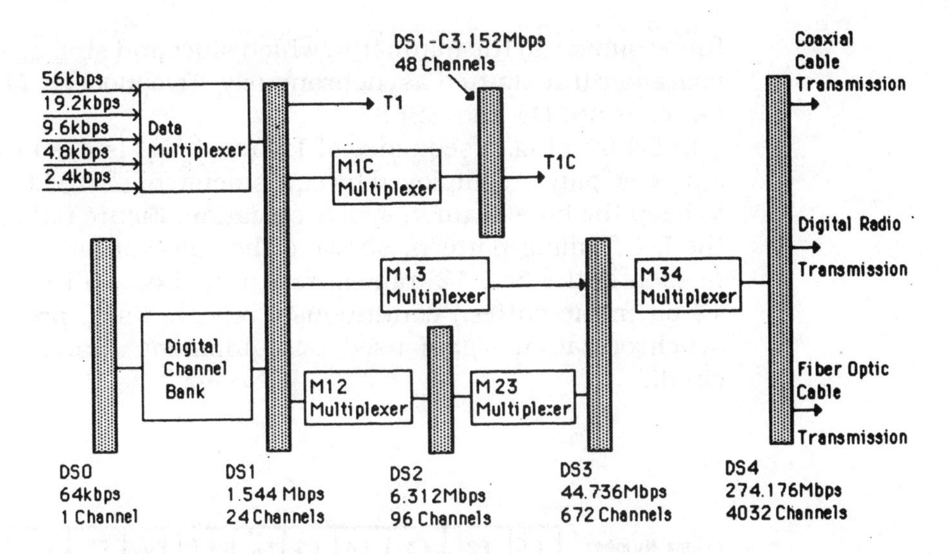

Figure 6.2 North American digital signal hierarchy.

channels operating at 6.312 Mbps. The AT&T M13 multiplexer operates upon 28 DS1 inputs, while that carrier's M23 multiplexer operates upon seven DS2 inputs, with both devices generating a DS3 output operating at 44.736 Mbps which represents 672 DS0 channels. The highest order AT&T multiplexer, which is the M34 device, accepts six DS3 inputs to form one DS4 output operating at 274.176 Mbps which represents 4032 DS0 channels.

Signaling restrictions

Data on a T1 circuit is transmitted using bipolar signaling, where binary zeros are represented by a 0 voltage, while binary ones are represented by alternating 3.0 volt positive and negative pulses. Thus, a second name for this signaling technique is alternate mark inversion (AMI).

There are several restrictions on the format of data transmitted on T1 circuits which multiplexers must follow. First, there can be no more than 15 consecutive binary zeros present in the data stream, requiring digital modems known as channel service units to insert and remove binary ones to keep the line in synchronization. Second, a framing pattern is required to provide synchronization since data on a T1 circuit is transmitted asynchronously. Here the framing pattern is used to indicate the beginning of one frame and the end of a preceding frame similar to the manner in which start and stop bits frame a character transmitted asynchronously. Two modern T1 framing patterns are D4 and ESF.

In D4 framing, a sequence of 12 frame bits is used to develop a precise pattern employed by equipment connected to a T1 line to keep the bit stream in synchronization. Figure 6.3 illustrates the D4 framing pattern, showing the value of bit number 193 in each of the first 12 frames transmitted on a T1 circuit. This 12-bit frame pattern continuously repeats itself, providing the synchronization signal used by equipment attached to a T1 circuit.

Frame Number	F1	F2	F3	F4	F5	F6	F7	F8	F9	F10	F11	F12
Frame Bit Value	1	0	0	0	1	1	0	1	1	1	0	0

Figure 6.3 The D4 framing pattern consists of the frame bit in each of 12 frames. This pattern repeats itself, providing equipment connected.

A more modern T1 framing pattern called the extended superframe (ESF) format was designed to provide performance information concerning the T1 line in addition to providing frame synchronization. This framing pattern contains 24 frame bits. Unlike the D4 pattern which repeats itself, the ESF consists of three types of frame bits, two of which can vary. These three types of frame bits include line control, cyclic redundancy checking, and a frame pattern for synchronization.

In the ESF frame, the bits in the odd frames from 1 through 23 are used by the telephone company to perform networking monitoring, set alarm conditions, and perform other control operations. The frame bits in frames 2, 6, 10, 14, 18, and 22 are used for cyclic redundancy checking. Finally, the frame bits in frames 4, 8, 12, 16, 20, and 24 are used to form the repeating pattern 001011, which is used for synchronization.

At the time this book was prepared, ESF was in use on approximately 25% of T1 circuits. As its implementation increases, users of T1 networks supporting this frame mechanism will obtain the ability to continuously check the performance of the network without having to inhibit traffic performance.

6.2 T-CARRIER MULTIPLEXERS

T-carrier multiplexers were originally point-to-point devices with each multiplexer considered to be a terminal or end-unit device that neither provided networking capability nor the ability to dynamically assign bandwidth utilization. Commensurate with the growth in T-carrier networking has been a corresponding increase in the features and capabilities of T-carrier multiplexers. Many T-carrier multiplexers now include multinodal support capability, the ability to perform many types of voice digitization through the addition of voice digitization modules as well as the ability to dynamically assign voice, data and video to the T-carrier bandwidth.

Since the primary difference between a T-carrier terminal or end-unit multiplexer and a T-carrier nodal switch is in the areas of multi-trunk support capability and the automatic rerouting of data, we will first focus our attention upon their common operational characteristics and features in this section. Using this information as a base will then enable us to examine the key differences between these two types of T-carrier multiplexers.

Operational characteristics

A basic T-carrier multiplexer operates in a manner similar to a high-capacity conventional TDM. That is, it accepts up to 24 or 30 DS0 inputs, each operating at 64 kbps, and builds a multiplexing frame in which the individual DS0's reside in fixed positions within the frame. For example, a 24-channel basic T1 multiplexer would build a frame by taking 8 bits from each input channel to construct a frame of 8 × 24, or 192 bits. The multiplexer then formats the frame into a bipolar AMI signal and passes the 192 bits to a channel service unit (CSU). The CSU terminates the customer end of the T1 line and provides an interface between the multiplexer and the communications carrier line.

CSU function

In addition to its line termination function, the CSU performs several additional functions. First, it adds a 193rd bit to each frame based upon the framing format supported by the communications carrier—D4 or ESF in the United States. Next, it is designed to recognize certain codes to initiate loopbacks as directed from a carrier's central office. If the CSU supports the ESF frame format, it will also keep statistics concerning the performance of the T1 line based upon the use of a CRC-6 in the ESF frame format. Figure 6.4 illustrates the relationship between a T1 multiplexer and a CSU.

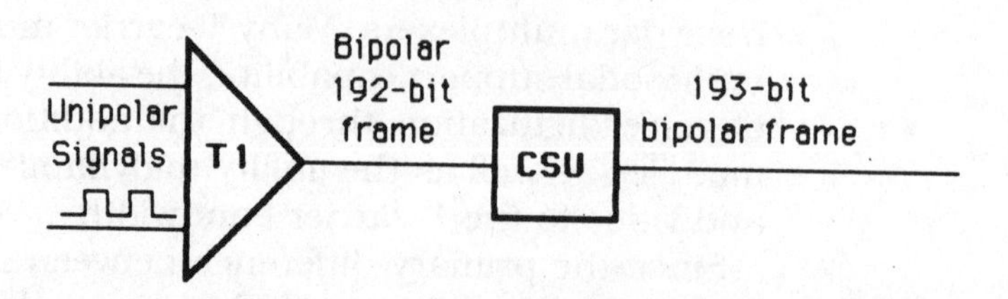

Figure 6.4 T1–CSU relationship. The T1 multiplexer constructs an 'unframed' 192-bit frame which is passed to the CSU. The CSU adds a frame bit to each frame based upon the framing format supported by the T1 line. In addition, the CSU is responsible for ensuring a minimum ones-density occurs on the line.

Although most CSUs are sold as stand-alone units, some T1 multiplexer vendors incorporate this device into their product as an option. In addition to the previously mentioned functions, the CSU is also responsible for ensuring that a minimum ones-density occurs on the line. The ones-density requirement relates to the use of metallic wire on span lines from the carrier's office to the customer's premises in which repeaters required at least one ones-pulse in every 15 bits to maintain timing.

The earliest method used to provide a minimum ones-density is known as binary 7 zero substitution in which the seventh bit in each byte is set to a one. While this method has a minimal effect on voice, it precludes using the full capacity of each DS0 for data and limits each DS0 to 56 kbps. A more modern and better method used to maintain a minimum ones-density is binary 8 zero substitution (B8ZS) in which an 8-bit byte containing all zero bits is encoded as a special bipolar violation containing two one bits. At the opposite end of the circuit, another B8ZS compliant CSU recognizes the 8-bit byte containing the bipolar violation and converts it back into an 8-bit byte containing all zero bits. Through the use of a B8ZS compatible CSU, each DS0 becomes capable of supporting a full 64 kbps operating rate. When this occurs, each DS0 is known as a clear channel DS0.

Application overview

In actuality, a basic T1 multiplexer can be considered as a D4 channel bank since essentially all T1 multiplexers add numerous features and capabilities which significantly extend the operational capability of the device. Later in this section we will consider a large number of features incorporated into most T1 multiplexers which extend their functionality and capability beyond that of a channel bank.

Some T-carrier multiplexers digitize voice directly through the addition of optional voice digitizer modules. Other T-carrier multiplexers require digitized voice to be routed as input to the multiplexer. Figure 6.5 illustrates a typical T-carrier multiplexer application where the device is used to combine a variety of digitized voice, data, and video inputs onto a T-carrier facility operating at a 1.544 Mbps (North American) or 2.048 Mbps (European) data rate.

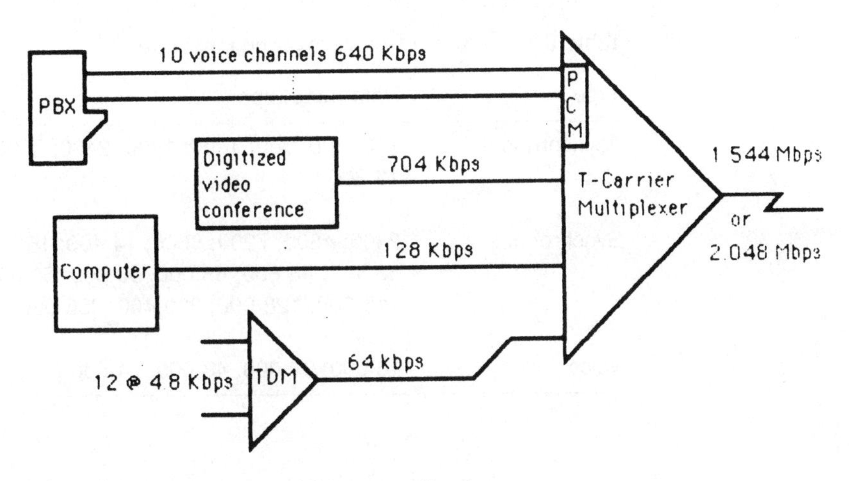

Figure 6.5 Typical T1 multiplexer application.

In the example illustrated in Figure 6.5, it was assumed that the digitized video conferencing required the use of 11 DS0 channels. Hence, the operating rate required to support full motion video through the multiplexer becomes 11 × 64 kbps, or 704 kbps. The 10 lines routed from the PBX are assumed to be analog, resulting in a requirement for voice digitizer modules to be installed in the T-carrier multiplexer. In this example, PCM digitization modules were used, resulting in each analog channel digitized at 64 kbps onto one DS0 channel. Thus, the 10 voice channels are shown as collectively occupying 640 kbps of the T-carrier bandwidth.

In the lower portion of Figure 6.5, 12 4.8 kbps data sources are first multiplexed by a conventional TDM, resulting in the composite 57.6 kbps data stream boosted to 64 kbps by the use of pad bits to enable its support by one T-carrier channel. The use of this type of multiplexer to pre-multiplex low data rate asynchronous or synchronous data sources will depend upon the functionality of the T-carrier multiplexer and its ability to multiplex subrate digital data streams. Some T-carrier multiplexers are limited to multiplexing synchronous data only. Other T-carrier multiplexers may support a variety of asynchronous and synchronous data rates through the use of different channel cards. Table 6.3 lists some of the more common data rates supported by T-carrier multiplexers, including the operating rate of many T-carrier voice digitization modules.

Table 6.3 Typical T1 multiplexer channel rates

Type	Data rates (bps)
Asynchronous	110; 300; 600; 1200; 1800; 2400; 3600; 4800; 7200; 9600; 19 200
Synchronous	2400; 4800; 7200; 9600; 14 400; 16 000; 19 200; 32 000; 38 400; 40 800; 48 000; 50 000; 56 000; 64 000; 112 000; 115 200; 128 000; 230 400; 256 000; 460 800
Voice	16 000; 32 000; 48 000; 64 000

Multiplexing efficiency

In addition to examining the type of voice, data, and video support it is also important to determine how efficiently a T-carrier multiplexer utilizes each DS0 channel. Some T-carrier multiplexers can only place one asynchronous or synchronous data source onto a DS0 channel regardless of its data rate; whereas, other multiplexers may make much more efficient utilization of DS0 channels. In our examination of T-carrier multiplexer features later in this chapter we will investigate the advantages of a subrate multiplexing capability.

6.3 FEATURES TO CONSIDER

The major features of T-carrier multiplexers that warrant consideration during an acquisition process are listed below. While all of the listed features are important to consider, they may not be relevant to certain situations based upon the immediate and long-term requirements of a specific organization.

- Bandwidth utilization method
- Bandwidth allocation method
- Voice interface support
- Voice digitization support
- Internodal trunk support
- Subrate channel utilization
- Digital access cross connect capability
- Gateway operation support
- Alternate routing and route generation
- Redundancy
- Maximum number of hops and nodes supported
- Diagnostics
- Configuration rules

Bandwidth utilization

Inefficient T-carrier multiplexers assign data to the T-carrier facility using 64 kbps DS0 channels for each data source as illustrated in Figure 6.6A. In this example the assignment of input data sources is fixed to predefined channels, resulting in the inability of the multiplexer to take advantage of the inactivity of different data sources.

More efficient T-carrier multiplexers employ a variety of demand-assigned bandwidth techniques to make more efficient use of the composite T-carrier bandwidth. This is illustrated in Figure 6.6B, in which a basic demand-assignment feature of a T-carrier multiplexer dynamically assigns bandwidth based upon the activity of the data sources. In this example it was assumed that several 9.6 kbps data sources became active along with two PCM digitized voice conversations and were dynamically assigned to the T-carrier bandwidth in their order of activation. Note that this method of bandwidth assignment normally results in an increase in available bandwidth since the probability of all inputs becoming active at one time is

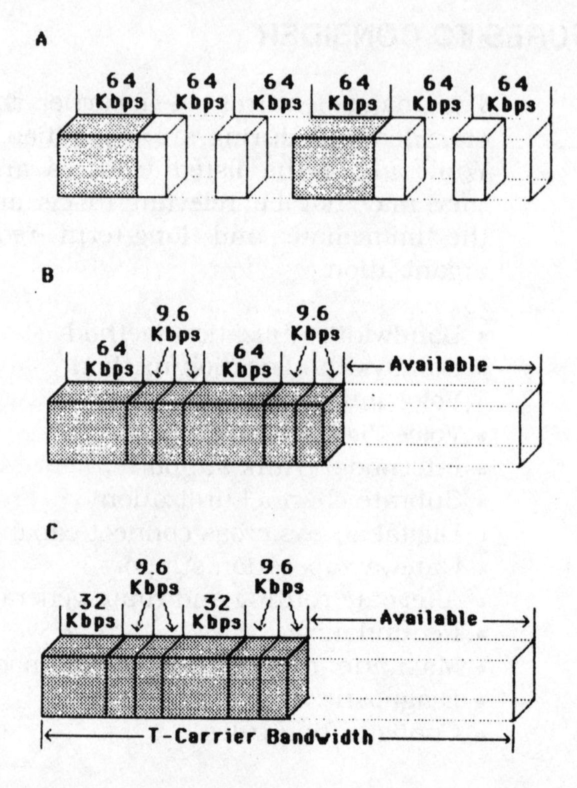

Figure 6.6 Demand-assigned bandwidth. A: conventional bandwidth allocation with PCM digitization. B: demand-assigned bandwidth with PCM digitization. C: demand-assigned bandwidth with ADPCM digitization.

usually very low. In addition, having the capability to allocate bandwidth based upon the data rate of the data sources and not by a DS0 channel basis allows the 9.6 kbps data sources to occupy significantly less bandwidth. Thus, demand assignment with dynamic bandwidth allocation results in a considerable improvement in the use of a T-carrier's data transmission capacity in comparison to a conventional bandwidth allocation process.

Figure 6.6C illustrates the effect upon bandwidth allocation based upon the use of a more efficient voice digitization module in T-carrier multiplexers. In this example it was assumed that ADPCM voice digitization modules were used in the T-carrier multiplexer instead of PCM voice digitization modules. The use of ADPCM reduces the bandwidth required for carrying each voice conversation to 32 kbps, further increasing the available bandwidth of the T-carrier to support other data sources.

Bandwidth allocation

Most T-carrier multiplexers use time division multiplexing schemes to allocate bandwidth to each voice and data channel as well as portions of DSO channels. Techniques used for bandwidth allocation can include the demand assignment of bandwidth previously illustrated in Figures 6.6B and C, as well as non-contiguous resource allocation and the packetization of voice, data, and video.

Figure 6.7 illustrates the advantage of non-contiguous resource allocation over the conventional method of allocating DSO channels. In Figure 6.7A a section of T-carrier bandwidth supporting three voice calls is shown. Here, each call is placed in a contiguous portion of the T-carrier bandwidth. In Figure 6.7B, it was assumed that call B was completed and its bandwidth became available for use. Now suppose a data source (D)

A. T-Carrier Bandwidth Section supporting three calls

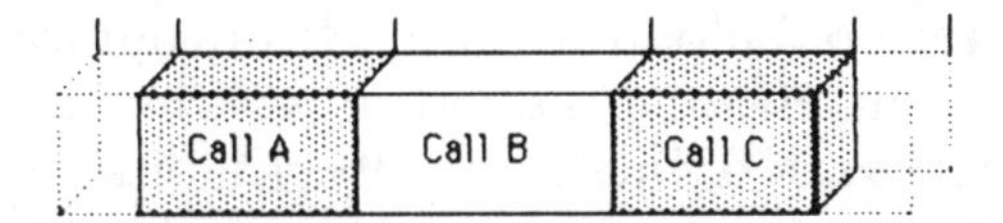

B. Call B is completed and its bandwidth becomes available

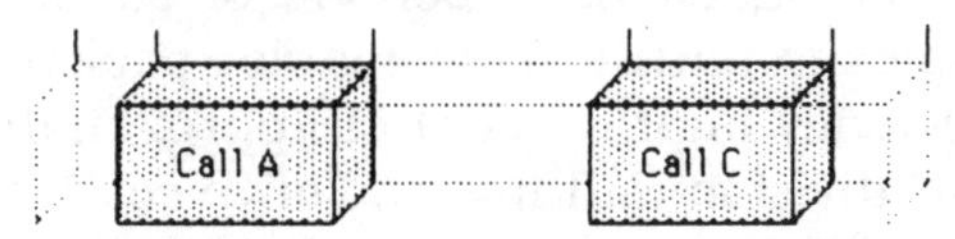

C. Data Source D multiplexed into non-contiguous sections of bandwidth

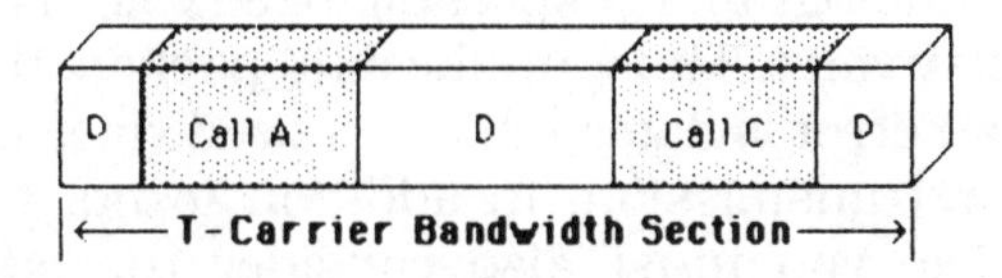

Figure 6.7 Bandwidth allocation methods. Non-contiguous resource allocation of bandwidth enables input to the T-carrier multiplexer to be split into portions of the available bandwidth. A: T-carrier bandwidth section supporting three calls. B: call B is completed and its bandwidth becomes available. C: data source D multiplexed into non-contiguous sections of bandwidth.

became active that required more bandwidth than that freed by the completion of call B. Under the non-contiguous resource allocation method the bandwidth required to accommodate the data source could be split into two or more non-contiguous sections of the T-carrier bandwidth as illustrated in Figure 6.7C.

The third method of bandwidth assignment was pioneered by Stratacom with that firm's introduction of a T-carrier multiplexer that packetizes both voice and data. The Stratacom multiplexer uses 'Fast Packet' technology where the term fast packet refers to the fact that information is transmitted across the network in packet format instead of a time division multiplexed format. Although the external interface to voice and data is the same as a conventional T-carrier multiplexer, the internal operation of the Stratacom multiplexer is considerably different from other devices.

Another interesting method of bandwidth allocation involves the packetization of voice and data sources onto T-carrier facilities. This technique was pioneered by Stratacom.

The Stratacom fast packet multiplexer generates packets only when data sources are active, using a packet length of 193 bits which corresponds to the North American T1 frame length. Figure 6.8 illustrates the Stratacom frame format. Although 24 bits, in effect, function as overhead to provide a destination address (16 bits), priority (2 bits), and error correction to the header by the use of a hamming code (6 bits), the efficiency of packetized multiplexing can be considerable. This is because the technique takes advantage of the fact that voice conversations have periods of silence and are typically half-duplex in nature. This enables packet technology to provide an efficiency improvement of approximately 2:1 over conventional time division multiplexing of voice. With the addition of ADPCM voice digitization modules the Stratacom fast packet multiplexer can support up to 96 voice conversations on a T1 circuit.

One of the problems associated with the use of packet technology to transport digitized voice is the fact that you cannot delay voice. Thus, unlike data packets that can be retransmitted if an error is detected, packetized voice cannot tolerate the delay of retransmission. In addition to not being able to retransmit voice, you must also consider the effect of a large number of voice channels becoming active. When too many channels become active the total bit rate of the digitized input channels can exceed the output bit rate of the T1 circuit. To avoid too much delay to specific channels, some channels will be skipped since a listener can tolerate a 125 ms delay. Another problem

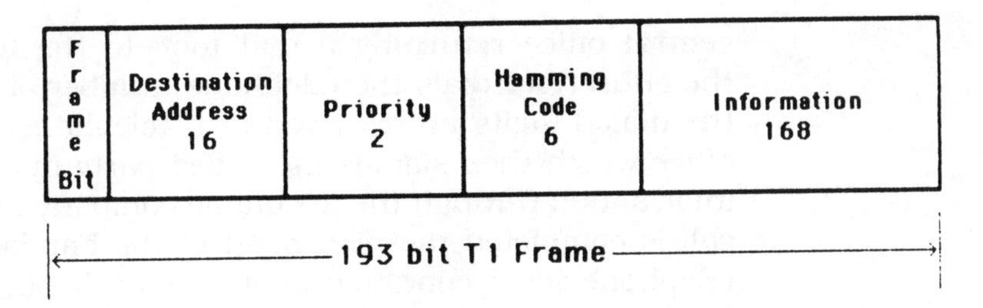

Figure 6.8 Stratacom packet format. The Stratacom T1 multiplexer packetizes voice and data sources into 193 bit frames containing 168 information bits.

associated with packetized voice is the delay that can occur as the packets are routed from node to node in a complex network. To overcome this problem Stratacom incorporates a priority field in its packet which enables certain packets to be processed and routed before other types of packets.

Voice interface support

Since most T1 multiplexer applications include the concentration of voice signals, the type of voice interfaces supported for two-wire and four-wire applications is an important multiplexer feature to consider. Prior to examining the types of voice interfaces supported by T1 multiplexers, let us first review some of the more common types of voice signaling methods since it is the signaling method that is actually supported by a particular interface.

Two of the most common types of telephone signaling include loop signaling and E&M signaling. Loop signaling is a signaling method employed on two-wire circuits between a telephone and a PBX or between a telephone and a central office. E&M signaling is a signaling method employed on both two-wire and four-wire circuits routed between telephone company switches.

Loop signaling

In loop signaling, the raising of the telephone handset results in the activation of a relay at the PBX or central office, causing current to flow in a circuit formed between the telephone set and the PBX or central office. The raising of the handset, referred to as an off-hook condition, results in the PBX or central office returning a dial tone to the telephone set. As the subscriber dials the telephone number of the called party, the dialed digits are received at a telephone company central office which then signals the called party by sending signaling information through the telephone company network. Once the call is completed the placement of the handset back onto the telephone set, a condition known as on-hook, causes the relay to be deactivated and the circuit previously formed to open.

A second type of telephone off-hook signaling that flows in a loop is ground start signaling. This method of signaling is also used on two-wire circuits between a telephone set and a PBX or central office. Unlike loop start signaling in which loop seizure is detected at the PBX or central office, ground start allows the detection of loop seizure to occur from either end of the line.

E&M signaling

E&M signaling is used on both two-wire and four-wire circuits connecting telephone company switches. Here the M lead is used to send ground or battery signals to the signaling circuits at a telephone company switch, while the E lead is used to receive an open or ground from the signaling circuit. In E&M signaling the local end asserts the M lead to seize control of the circuit. The remote end receives the signal on the E lead and toggles its M lead as a signal for the local end to proceed. The local end then sends the addrcss by toggling its M lead, in effect, placing dialing pulses on that lead which is used by the remote end to effect the desired connection. Once a call is completed, either party will drop its M lead, resulting in the other side responding by dropping its M lead. Currently, there are three types of E&M signaling: types I, II, and III. The difference between E&M signaling types relates to the method by which an on-hook condition is established—ground or open.

Some of the more common types of voice interface cards supported by many T-carrier multiplexer vendors are:

- Two-wire transmission only
- Two-wire E&M
- Two-wire foreign exchange
- Four-wire transmission only
- Four-wire E&M

The two-wire and four-wire transmission-only interfaces are designed to support permanent two-wire and four-wire connections between two points that do not require the passing of signaling information. Both types of interfaces are normally used to support a data modem connection through a T-carrier multiplexer.

The two-wire and four-wire E&M interfaces usually support the connection of PBXs and telephone company equipment to a T-carrier multiplexer. As previously mentioned, there are three types of E&M signaling, with each type applicable to both two-wire and four-wire operations.

The two-wire foreign exchange office interface is designed to support the attachment of a T-carrier multiplexer to a PBX or central office switching equipment that provides an open or closed foreign exchange termination point.

Voice digitization support

In addition to the method of bandwidth assignment and allocation, a third major feature affecting the efficiency of a T-carrier multiplexer is the type of voice digitization modules supported by the device.

PCM

In a conventional PCM digitization process the height of the analog signal is converted into an 8-bit word which represents the analog signal at the time sampling occurred. Since sampling occurs 8000 times per second, an analog voice signal is converted into a 64 kbps digital data stream. To increase the number of voice signals that can be carried on a T1 or E1 transmission facility, a variety of voice digitization techniques have been developed, including adaptive differential pulse code modulation (ADPCM) and continuous variable slope delta modulation (CVSD), as well as several less widely employed schemes known as time assigned speed interpolation (TASI) and differential PCM (DPCM).

ADPCM

When adaptive differential pulse code modulation is employed, a transcoder is utilized to reduce the 8-bit samples normally associated with PCM into 4-bit words, retaining the 8000 samples per second PCM sampling rate. This technique results in a voice digitization rate of 32 kbps, which is one-half the PCM voice digitization data rate.

Under the ADPCM technique, the use of 4-bit words permits only 15 quantizing levels; however, instead of representing the height of the analog signal, each word contains information required to reconstruct the signal. This information is obtained by circuitry in the transcoder which adaptively predicts the value of the next signal based upon the signal level of the previous sample. This technique is known as adaptive prediction and its accuracy is based upon the fact that the human voice does not significantly change from one sampling interval to the next. Until 1988 there were no standard ADPCM algorithms and most T1 multiplexer vendors offering such modules employed proprietary transcoder algorithms. As a result of this lack of ADPCM algorithm standardization, some T1 multiplexers from one vendor may be incapable of directly passing through digitized voice into a T1 multiplexer manufactured by a different vendor, requiring ADPCM data channels to be first reconverted to voice and then redigitized prior to remultiplexing. Fortunately, the gradual adoption of the ADPCM standard by multiplexing vendors should eventually eliminate this incompatibility problem.

CVSDM

In the continuously variable slope delta modulation digitization technique, the analog input voltage is compared to a reference voltage. If the input is greater than the reference a binary one is encoded, while a binary zero is encoded if the input voltage is less than the reference level. This permits a 1-bit data word to represent each sample.

At the receiver, the incoming bit stream represents changes to the reference voltage and is used to reconstruct the original analog signal. Each one bit causes the receiver to add height to the reconstructed analog signal, while each zero bit causes the receiver to decrease the analog signal by a set amount. If the reconstructed signal is plotted, the incremental increases and decreases in the height of the signal result in a series of changing slopes, resulting in the naming of this technique—continuously variable slope delta modulation.

Since only changes in the slope or steepness of the analog signal are transmitted, a sampling rate higher than the PCM sampling rate is required to recognize rapidly changing signals. Typically, CVSD samples the analog input at 16 000 or 32 000 times per second. With a 1-bit word transmitted for each sample, the CVSD data rate normally is 16 kbps or 32 kbps. Other CVSD data rates are obtainable by varying the sampling rate. Some T1 multiplexer manufacturers offer a CVSD option which permits sampling rates from 9600 to 64 000 samples per second, resulting in a CVSD data rate ranging from 9.6 kbps to 64 kbps, with the lower sampling rates reducing the quality of the reconstructed voice signal. Normally, voice signals are well recognized at 16 kbps and above, while a data rate of 9.6 kbps will result in a marginally recognizable reconstructed voice signal.

Although most T-carrier multiplexers support the use of PCM and ADPCM, some vendors also support the use of adapter cards that contain proprietary voice digitization modules. One example of this is adaptive speech interpolation which changes the digitization rate of selected voice channels from 32 kbps to 24 kbps as available bandwidth becomes saturated. Although some proprietary techniques may offer advantages in both the fidelity of a reconstructed voice signal as well as in the bandwidth required to carry the signal, their use restricts an organization to one vendor's product.

Figure 6.9 illustrates the effect of the use of several types of voice digitization modules upon the capacity of a T-carrier. If standard PCM modules are used, the T-carrier becomes capable of supporting either 24 or 30 voice calls depending upon whether a North American or European T-carrier facility is used. When 32 kbps ADPCM modules are used to digitize voice, the voice-carrying capacity of the T-carrier is doubled as shown in Figure 6.9B. Figure 6.9C, which illustrates the use of 24 kbps ADPCM, shows the voice-carrying capacity of a T-carrier tripling, while Figure 6.9D shows how the voice-carrying capacity of a T-carrier can be quadrupled through the use of ADPCM and DSI.

Although the utilization of such voice digitization techniques as ADPCM and CVSDM can increase the channel capacity of a T-carrier multiplexer and its ability to service additional voice conversations, this extra capacity may impede the data-carrying capacity of the multiplexer. This is because ADPCM is limited to carrying modem signals operating at or under

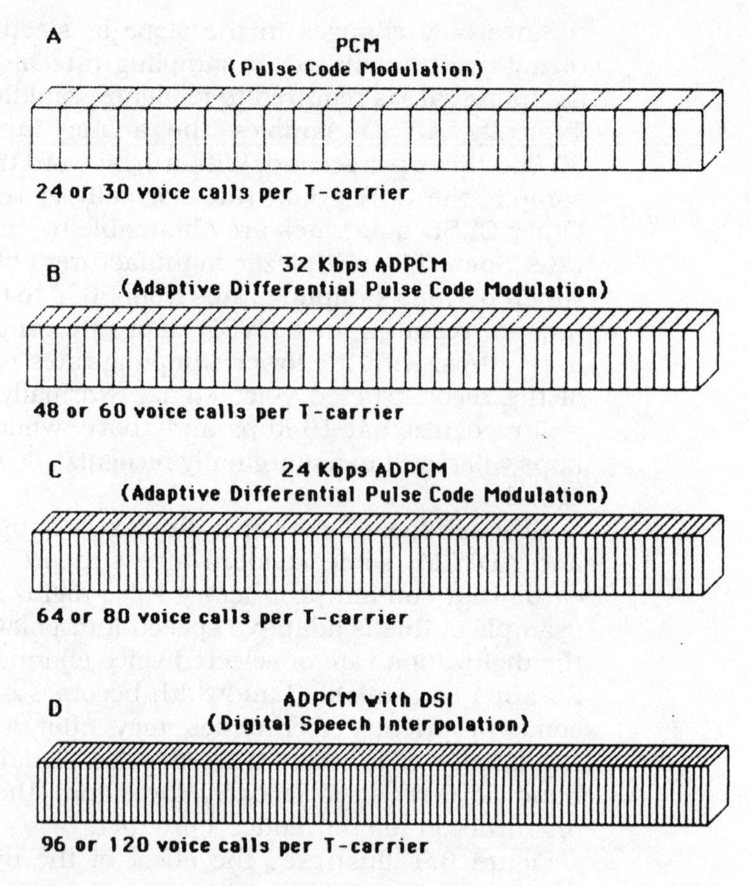

Figure 6.9 Trunk capacity as a function of digitization.

4800 bps, while other voice digitization techniques that produce speech at a rate at or below 24 kbps are not recommended for carrying modem data. This restriction means that as you increase the voice-carrying capacity of a T-carrier multiplexer through the use of voice digitization techniques other than PCM you simultaneously reduce the capability of the multiplexer with respect to passing dial-up and leased line connections used by modems.

Internodal trunk support

The internodal trunk support feature of T-carrier multiplexers describes the ability of the device to connect to North American and European T-carrier facilities. To support North American T-carrier facility usage, the multiplexer must not only operate at 1.544 Mbps but, in addition, support the required communications carrier framing—D4 or ESF. To support European T-carrier facility usage the multiplexer must operate at 2.048 Mbps and support CEPT 30-PCM framing.

Subrate channel utilization

Channel utilization is a function of the subrate multiplexing capabilities of the T-carrier multiplexer. Many T-carrier multiplexers support asynchronous data rates from 50 bps to 19.2 kbps and synchronous data rates from 2.4 kbps to 19.2 kbps, permitting multiple data sources to be placed onto one DSO channel. Figure 6.10 illustrates one of the methods by which a T-carrier multiplexer vendor's equipment might multiplex subrate data channels onto one DSO channel. Unfortunately, not all vendors provide a subrate multiplexing capability in their equipment. When this occurs, subrate data sources are bit padded to operate at 64 kbps which can considerably reduce the ability of the multiplexer to maximize bandwidth utilization. In such situations users can obtain one or more subrate multiplexers to combine several data sources to a 64 kbps data rate; however, this may result in a higher cost than obtaining T-carrier multiplexers that include a built-in subrate multiplexing capability.

<table>
<tr><th colspan="12">ONE DSO CHANNEL</th></tr>
<tr><td colspan="4">19.2</td><td colspan="4">19.2</td><td colspan="4">19.2</td></tr>
<tr><td colspan="4">16.8</td><td colspan="4">16.8</td><td colspan="4">16.8</td></tr>
<tr><td colspan="2">9.6</td><td colspan="2">9.6</td><td colspan="2">9.6</td><td colspan="2">9.6</td><td colspan="2">9.6</td><td colspan="2">9.6</td></tr>
<tr><td>4.8</td><td>4.8</td><td>4.8</td><td>4.8</td><td>4.8</td><td>4.8</td><td>4.8</td><td>4.8</td><td>4.8</td><td>4.8</td><td>4.8</td><td>4.8</td></tr>
<tr><td>2.4</td><td>2.4</td><td>2.4</td><td>2.4</td><td>2.4</td><td>2.4</td><td>2.4</td><td>2.4</td><td>2.4</td><td>2.4</td><td>2.4</td><td>2.4</td></tr>
<tr><td>1.2</td><td>1.2</td><td>1.2</td><td>1.2</td><td>1.2</td><td>1.2</td><td>1.2</td><td>1.2</td><td>1.2</td><td>1.2</td><td>1.2</td><td>1.2</td></tr>
<tr><td>1.2</td><td>1.2</td><td>2.4</td><td>2.4</td><td>4.8</td><td>4.8</td><td colspan="2">9.6</td><td colspan="4">19.2</td></tr>
</table>

Figure 6.10 Typical subrate channel utilization.

Digital access cross connect capability

The ability of a multiplexer to provide digital access cross connect (DAC) operations can be viewed as the next step up on terms of functionality from point-to-point multiplexer operations. Although communications carrier DACs are limited to switching DS0 channels, many multiplexer vendors include the capability to drop and insert/bypass subrate channels or digitized voice encoded at bit rates under 64 kbps, permitting sophisticated T-carrier networks to be constructed.

Figure 6.11 illustrates an example of the use of a digital access cross connect feature used in three T-carrier multiplexers labeled A, B, and C. In this example, channel 8 on multiplexer A is routed to multiplexer C where it is dropped, freeing that DS0 channel for use as the T-carrier is then routed to multiplexer B's location. Thus, a data source at location C could be inserted into the T-carrier on channel 8, resulting in channel 8 being routed from C to B as shown in Figure 6.11. In this example it was assumed that all other DS0 channels were simply passed through or bypass multiplexer C and are then routed to multiplexer B. Thus, a digital access and cross connect capability can be used to establish a virtual circuit through an intermediate multiplexer (bypass) without demultiplexing the data, to allow intermediate nodes to add data to the data stream (insert), as well as to permit an intermediate node to act as a terminating node (drop) for other multiplexers.

One limitation associated with the use of T-carrier multiplexers is the ability of the multiplexer to be used with a communications carrier's digital access cross connect system. For example, you may want to install a T1 multiplexer at one of your organization's locations and have several DS0s dropped at the carrier's central office, where they are connected to a specific type of communications facility, such

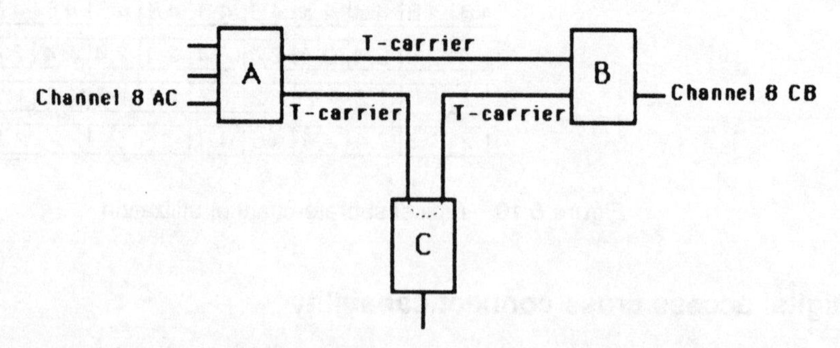

Figure 6.11 DS0 cross connect. Although many multiplexers support the cross connection of DS0 channels, some T-carrier multiplexers also permit the cross connection of subchannels.

as an interstate or intrastate WATS or FX line connection. To accomplish this your T-carrier multiplexer must operate as a byte interleaved multiplexer in comparison to operating as a bit interleaved multiplexer. Doing so enables the communications carrier's equipment to identify and extract the appropriate DS0s from the framed DS1 signal.

Gateway operation support

To function as a gateway requires a T-carrier multiplexer to support a minimum of two high-speed circuits. In addition, the T-carrier multiplexer must perform several other operations that must be coordinated with the use of the T-carrier multiplexers connected to the gateway multiplexer.

Three of the main problems associated with connecting European and North American T-carrier circuits through the use of a gateway multiplexer involve compensating for the differences between European and North American T-carriers with respect to the number of DS0 channels each T-carrier supports, the method by which signaling is carried in each channel, and the method by which performance monitoring is accomplished.

Since a European T-carrier contains 30 DS0 channels, while a North American T1 link supports 24, the gateway multiplexer will map 30 DS0 channels to 24, resulting in the loss of six channels. This means that the end-user's organization is limited to the effective use of 24 channels on a European connection via a gateway multiplexer.

For signaling conversion the gateway multiplexer will move AB or ABCD signaling under D4 and ESF frame formats into channel 16 for North American to European conversion. For signaling conversion in the other direction, appropriate bits will be moved from channel 16 to the robbed bit positions used in D4 and ESF framing.

The area of performance monitoring is presently either ignored by gateway multiplexers or handled on an individual link basis to the gateway device. Since European systems use a CRC-4 check while ESF employs a CRC-6 check no conversion is performed by multiplexers that support performance monitoring since the results would be meaningless for a link consisting of both North American and European facilities connected through a gateway. Instead, the gateway will provide statistics treating each connection as a separate

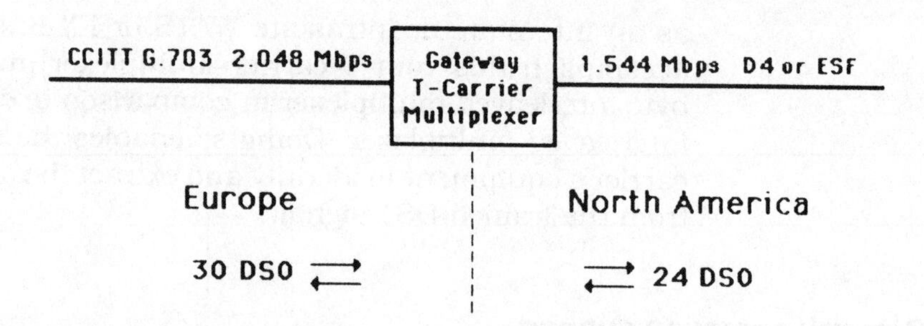

Figure 6.12 Gateway operation. The operation of a gateway results in the dropping of six DS0 channels when a European CEPT-30 facility is converted to a North American T1 link.

T-carrier facility. Figure 6.12 illustrates the placement of a gateway T-carrier multiplexer connecting a North American T1 facility to a European CEPT-30 facility.

Alternate routing and route generation

If a T-carrier network consists of three or more multiplexers interconnected by those carrier facilities, both the alternate routing capability and the method of route generation are important features to consider.

Basically, route generation falls into two broad areas: paths initiated by tables constructed by operators and dynamic paths automatically generated and maintained by the multiplexers. To illustrate both alternate routing and route generation, consider the T-carrier network illustrated in Figure 6.13. In this illustration the T-carrier circuit connecting multiplexers A and B has failed. With an alternate routing capability some or all DS0 channels previously carried by circuit AB must be routed from path AC to path CB to multiplexer B.

If the multiplexers employ alternate routing based upon predefined tables assigned by operations personnel, DS0 channels previously routed on path AB will be inserted into the T-carrier linking A to C and then routed onto path CB based upon the use of those tables. As the DS0 channels from path AB are inserted into the T-carrier linking A to C, DS0 channels on path AC must be dropped, a process referred to as bumping. If calls were in progress on path A to C and C to B when the failure between A and B occurred, those calls may be dropped depending upon whether or not the multiplexers

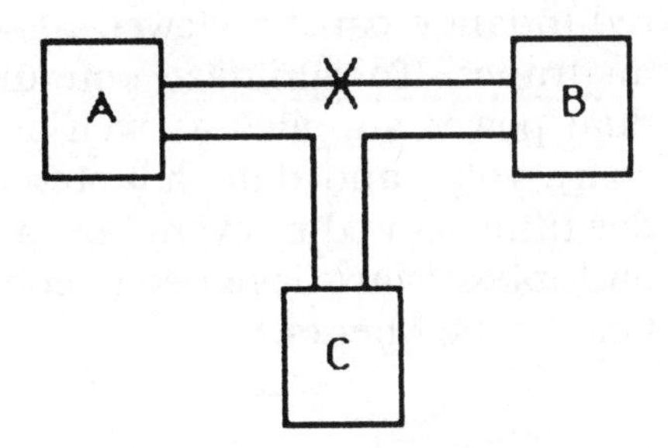

Figure 6.13 Alternate routing. T-carrier multiplexers provide a variety of methods to effect alternate routing, including using predefined tables and by the dynamic examination of current activity at the time of failure.

employ a priority bumping feature or have the capability to downspeed voice digitized DS0 channels.

Priority bumping refers to the ability to override certain existing DS0 subchannels based upon the priority assigned to DS0 channels that were previously carried on the failed link and the priorities assigned to DS0 active channels on the operational links. Downspeed refers to the capability of multiplexers to shift to a different and more efficient voice digitization algorithm to obtain additional bandwidth to obtain the ability to carry DS0 channels from the failed link on the operational circuits. One example of downspeed would be switching from 32 kbps ADPCM to 24 kbps ADPCM, resulting in freeing up 12 kbps per operational DS0 channel.

When alternate routing and route generation is dynamically performed by multiplexers, those devices examine current DS0 activity and establish alternate routing based upon predefined priorities and the current activity of DS0 channels. If predefined tables are used, the multiplexers do not examine whether or not a particular DS0 channel is active prior to performing alternate routing; however, some multiplexers may have the ability to perform forced or transparent bumping regardless of whether they use fixed tables or dynamically generate alternate paths. Under forced bumping, DS0 channels are immediately reassigned, whereas under transparent bumping current voice or data sessions are allowed to complete prior to their bandwidth being reassigned.

Redundancy

Since the failure of a T-carrier multiplexer can result in a large number of voice and data circuits becoming inoperative, redundancy can be viewed as a necessity similar to business insurance. To minimize potential downtime you can consider dual power supplies as well as redundant common logic and spare voice and data adapter cards. Doing so may minimize downtime in the event of a component failure as many multiplexers are designed to enable technicians to easily replace failed components.

Maximum number of hops and nodes supported

As T-carrier multiplexers are interconnected to form a network each multiplexer can be considered as a network node. When a DS0 channel is routed through a multiplexer that multiplexer is known as a hop. Thus, the maximum number of hops refers to the maximum number of internodal devices a DS0 channel can traverse to complete an end-to-end connection.

In addition to the maximum number of hops users must also consider the maximum number of nodes that can be networked together. The maximum number of addressable nodes that can be managed as a single network is normally much greater than the maximum number of hops supported since the latter is constrained by the delay to voice as DS0 channels are switched and routed through hops.

Diagnostics

Most T-carrier multiplexers provide both local and remote channel loop-back capability to facilitate fault isolation. Some multiplexers have built-in test pattern generation capability which may alleviate the necessity of obtaining additional test equipment for isolating network faults.

Configuration rules

Figure 6.14 illustrates a typical T-carrier multiplexer cabinet layout which is similar to the manner in which a multiplexer would be installed in an industry standard 19-inch rack. In examining multiplexer configuration rules, a variety of constraints may exist to include the number of trunk module cards, voice cards, and data cards that can be installed. Other constraints will include the physical number of channels

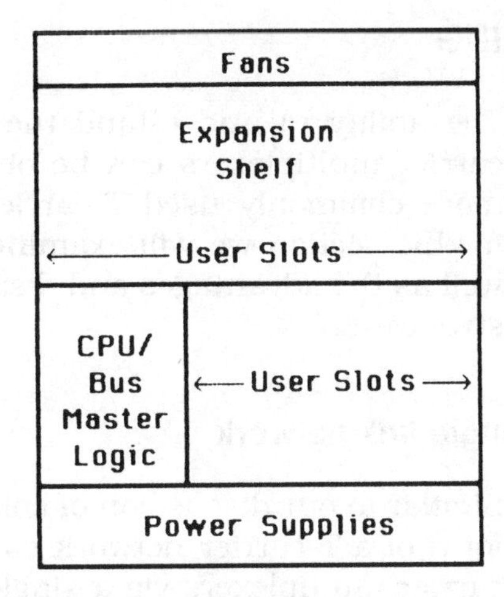

Figure 6.14 Typical T-carrier multiplexer cabinet, layout.

supported by each card and the type of voice digitization modules that can be obtained. Depending upon end-user requirements, additional expansion shelves may be required to support additional cards. When this occurs, additional power supplies may be required and their cost and space requirements must be considered.

Multiplexers and nodal processors

Due to the liberty by which vendors can label products, there is no definitive line that separates a T-carrier terminal or end-unit multiplexer from a nodal processor. In general, we can categorize each device by the number of trunks they support, the method by which alternate routing is performed, and the method by which operators control the network.

In general, a nodal processor supports more than 16 T-carrier circuits and includes the capability to dynamically perform alternate routing based upon one or more algorithms. In addition, this device normally permits network configuration to be effected from a central node. In comparison, a T-carrier terminal or end-unit multiplexer supports up to 16 trunks and usually relies upon the use of predefined tables to perform alternate routing, assuming they actually have this capability. In addition, network reconfiguration may require operators to reprogram each multiplexer individually.

6.4 NETWORKING

The ability to understand the versatility and capability of T-carrier multiplexers can be obtained from an overview of the more commonly used T-carrier multiplexer topologies. Thus, in this section we will examine several network structures as well as the advantages and disadvantages associated with each structure.

Point-to-point single link network

Similar to our discussion of conventional TDMs, the most basic form of a T-carrier network consists of interconnecting two T-carrier multiplexers via a single T1 or E1 line, resulting in the establishment of a point-to-point network structure.

Figure 6.15 illustrates the utilization of T-carrier multiplexers in a point-to-point networking environment. Although this type of network probably represents the most common use of T-carrier multiplexers, the single T1 or E1 circuit used to interconnect the multiplexers will result in a line failure causing the entire network to become inoperative. Due to this, many organizations may prefer to use multi-node T-carrier multiplexers that are interconnected by multiple T1 or E1 circuits.

Point-to-point multiple link network

Through the use of multi-node T-carrier multiplexers you can interconnect two locations through the use of multiple T1 or E1

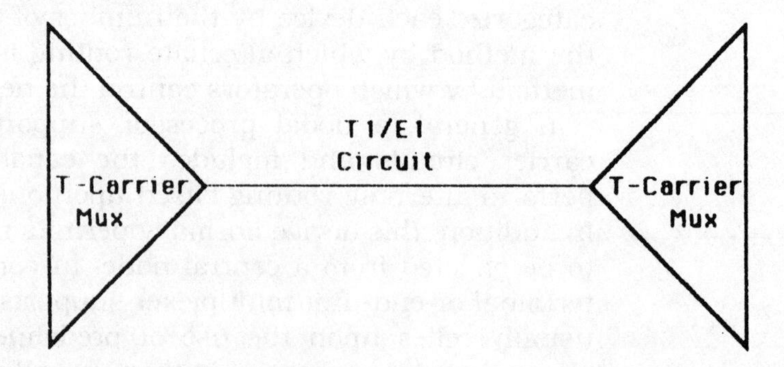

Figure 6.15 Point-to-point single link network. In this type of network a link failure renders the entire network inoperative.

circuits. Although one circuit can serve as backup to the second circuit, to ensure the backup circuit is available when needed you must order the circuits to be routed through different communications carrier central offices. To do so, you would order diversity routing which, for a small additional fee, requires the communications carrier to ensure that each T1 or E1 circuit is routed through different paths, different higher order microwave or fiber optic transmission facilities, and different carrier offices. Doing so ensures that a backhoe operator that inadvertently cuts a fiber cable or the occurrence of a fire in a carrier's central office does not affect both lines.

Figure 6.16 illustrates a point-to-point multiple link network configuration. As an alternative to simply using one circuit just for backup, many organizations will use both circuits in a load-sharing configuration. Doing so enables 48 or 60 DS0s to be supported when both circuits are operational. Then, if one T1 or E1 circuit should fail priority bumping and/or downspeeding could be used to continue a majority of the transmission between the two locations on the remaining circuit.

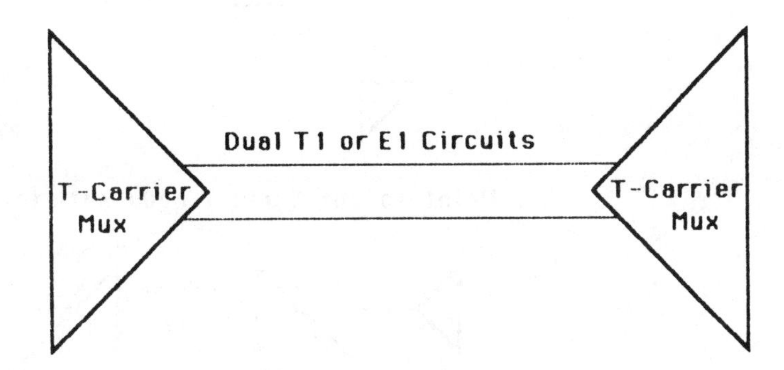

Figure 6.16 Point-to-point multiple link network. When multiple T1 or E1 circuits are used to interconnect T-carrier multiplexers, the failure of one link can be compensated for by transmission on the operational circuit.

Star network

For organizations that have a large headquarters or central site facility and many distributed offices, a star network configuration should be considered. Under this topological arrangement each distributed office is connected to the headquarters office or central site via a point-to-point link.

At the headquarters location, either individual T-carrier multiplexers or multi-nodal multiplexers can be used to service the transmission requirements of each remote location. If individual T-carrier multiplexers are used at the headquarters office or central site to service individual remote offices, this will normally preclude the ability of one remote location to transfer voice and/or data to another remote location. By installing multi-nodal T-carrier multiplexers you can obtain the ability to provide remote offices with the capability to communicate with other remote offices through equipment installed at the central site.

Figure 6.17 illustrates two types of star network topologies. In Figure 6.17A each remote location is conected to the central site

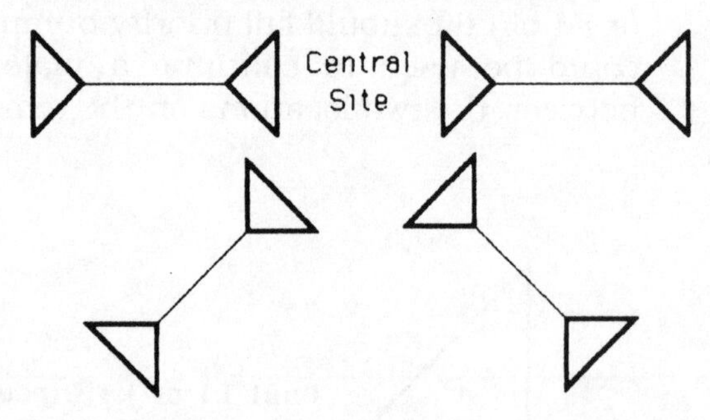

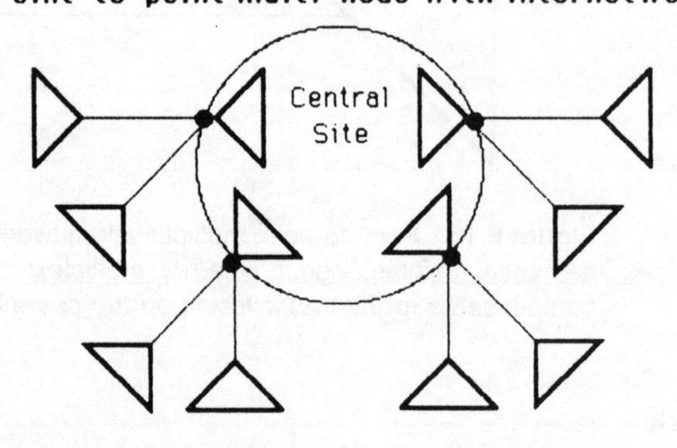

Figure 6.17 Star network configurations.

via a point-to-point link, with individual T-carrier multiplexers used to service each remote office. In Figure 6.17B illustrates the use of multi-nodal T-carrier multiplexers at the central site; in addition to each multi-nodal multiplexer supporting two or more remote offices the multi-nodal multiplexers can be interconnected, providing the ability for a data source residing on a channel of a T-carrier multiplexer at one remote location being cross connected and routed to a channel destined to a T-carrier multiplexer at another remote location. In this example the T-carrier multiplexers at the central site are shown interconnected via T1 or E1 loops which form a ring. While this ring structure is illustrated for a local interconnection of T-carrier multiplexers, you can also interconnect remote locations to one another via a ring structure.

Although single node T-carrier multiplexers are illustrated in Figure 6.17A, multi-node multiplexers can also be used to provide multiple links between each remote location and the central site. Doing so would provide a higher level of reliability; however, unless the central site multiplexers are interconnected as illustrated in Figure 6.17B, you would not obtain an internetworking capability.

Ring network

In a ring network structure, T-carrier multiplexers are interconnected to form a ring as illustrated in Figure 6.18. In addition to providing an internetworking capability, when properly configured for alternate routing a ring structure can be used to provide a high degree of circuit redundancy. For example, the failure of a T1 or E1 circuit connecting two sites can be compensated for by routing traffic around the ring in the reverse direction. Of course, bumping and/or channel downspeeding would play an important role to ensure priority traffic sent in a reverse direction is received at its destination.

Multipoint network

Similar to analog modem and Dataphone Digital Service networking options you can route a T1 or E1 circuit between three or more locations, forming a multipoint or multidrop network structure. Figure 6.19 illustrates a multipoint T-carrier network structure in which three locations are interconnected via the use of two T1 or E1 circuits.

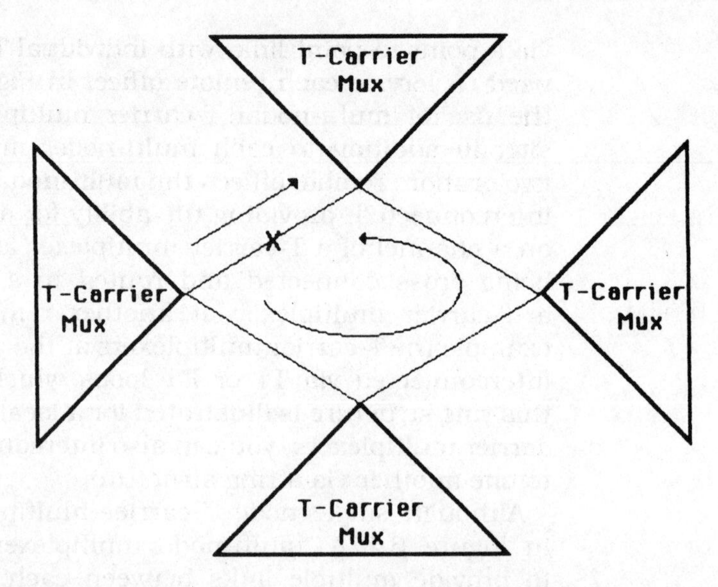

Figure 6.18 In a ring network, the failure of a link is partially compensated for by rerouting priority traffic around the ring in the reverse direction.

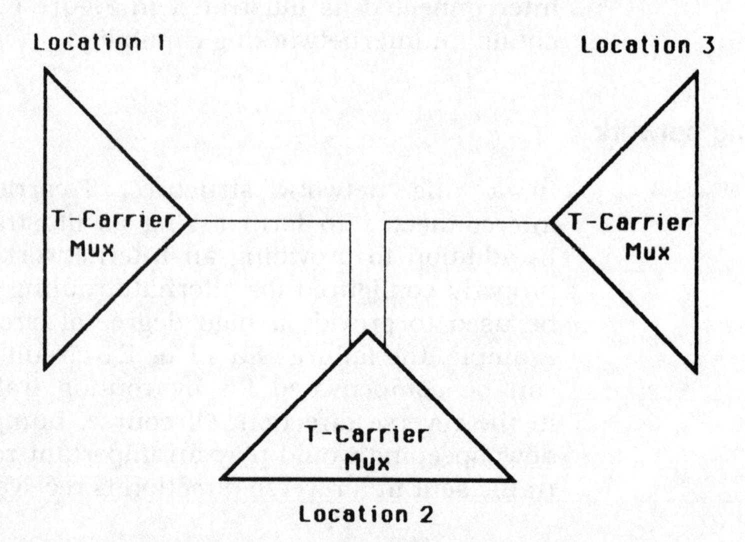

Figure 6.19 Multipoint network. When a multipoint network structure is used, T-carrier multiplexers at intermediate nodes must be capable of performing drop, insert, and pass-thru operations.

The effective utilization of a multipoint network structure is based upon only a portion of each T1 or E1 line segment being used to interconnect two locations, enabling the excessive capacity of a line segment to be used to carry traffic to a distant node. In the example illustrated in Figure 6.19, the intermediate node at location 2 must be capable of performing drop, insert and pass-thru operations to remove channels from location 1 routed to location 2, pass channels from location 1 routed to location 3, and insert channels from location 2 routed to locations 1 and 3.

Mesh network

In a T-carrier mesh network, each T-carrier multiplexer is connected to every other multiplexer which results in a mesh topology. Figure 6.20 illustrates a T-carrier mesh network interconnecting four T-carrier multiplexers.

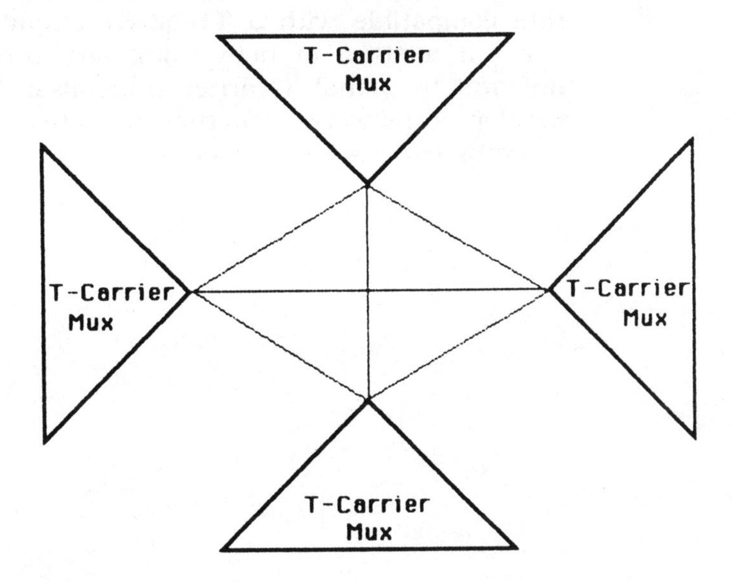

Figure 6.20 Mesh network. Each T-carrier multiplexer is connected to every multiplexer in the network.

Although a mesh structure provides the highest degree of redundancy available from any network structure, it is also the most expensive to implement. This is because the mesh structure requires the use of more T1 or E1 circuits than other networking structures. In spite of its high cost, many organizations will consider a mesh network structure for a portion of their entire backbone network. In doing so, the mesh network would probably be used to interconnect major offices that have a high level of data traffic since this network structure results in a minimum transit time for traffic flow when the T-carrier multiplexers support load balancing.

Due to the expenses associated with T-carrier mesh networks, organizations that do not require a full T-carrier between locations but whose requirements are greater than a single DS0 have begun to use fractional T1 and fractional E1 transmission facilities. Since access to a fractional T1 or E1 circuit is via a full T1 or E1 circuit to a carrier's central office, both T-carrier multiplexers as well as specialized fractional T1/E1 multiplexers can be used to access a carrier's fractional T1/E1 transmission facility. Concerning specialized fractional T1/E1 multiplexers, such products can be considered as a limited-function T-carrier multiplexer as they accept only a few DS0 inputs although they produce a composite data rate compatible with a T1 or E1 circuit operating rate. By using a mixture of networking structures as well as both full and fractional T-carrier transmission facilities, you can develop a network structure to satisfy your organization's communications requirements.

7

FIBER OPTIC MULTIPLEXERS

While the majority of attention focused upon fiber optic transmission relates to its use by communications carriers, this technology can be directly applicable to many corporate networks. Due to the properties of light transmission, fiber optic systems are well suited for many specialized applications, including the high-speed transmission of data between terminals and a computer or between computers located in the same building.

Similar to the primary rationale for the use of other types of multiplexers, fiber optic multiplexers are designed to enable many data sources to share the use of a common communications circuit. However, instead of sharing the use of a conventional twisted-pair-based media, the fiber optic multiplexer enables the use of a fiber optic cable to be shared.

Although fiber optic multiplexers operate in a manner similar to time division multiplexers, the ability of many devices to directly interface fiber optic cable results in some key differences between multiplexers. These differences include the inclusion of electrical-to-optical and optical-to-electrical converters in many fiber optic multiplexers as well as their ability to support a larger mixture of high data rates due to the larger bandwidth of fiber in comparison to conventional metallic cable. Since many of the advantages of fiber optic multiplexers are related to the use of fiber optic cable, we will first focus our attention upon the components of a fiber optic transmission system, including the transmitter, transmission medium, and receiver used with optical transmission. After examining the advantages and limitations associated with such systems we will use the information previously presented as a foundation to analyze the economics associated with cabling terminals to a computer, illustrating how this technology can be directly applicable to many computer installations. In doing so, we will examine the use of fiber optic modems and multiplexers to denote the advantages and limitations associated with their use in a communications network.

7.1 SYSTEM COMPONENTS

Similar to conventional transmission systems, the major components of a fiber optic system include a transmitter, transmission medium, and receiver. The transmitter employed in a fiber optic system is an electrical-to-optical (E/O) converter. The E/O converter receives electronic signals and converts those signals into a series of light pulses. The transmission medium is an optical fiber cable which can be constructed out of plastic or glass material. The receiver used in fiber optic systems is an optical-to-electrical (O/E) converter. The O/E converter changes the received light signals into their equivalent electrical signals. The relationship of the three major fiber optic system components is shown in Figure 7.1.

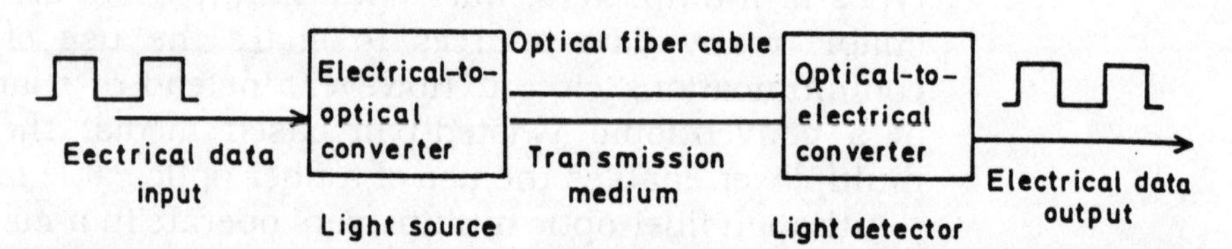

Figure 7.1 Fiber optic system component relationship. The electrical-to-optical converter produces a light source which is transmitted over the optical fiber cable. At the opposite end of the cable an optical-to-electrical converter changes the light signal back into an electrical signal.

The light source

Two types of light sources dominate the electronic-to-optical conversion market: the light emitting diode (LED) and the laser diode (LD). Although both devices provide a mechanism for the conversion of electrically encoded information into light encoded information, their utilization criteria depends upon many factors. These factors include response times, temperature sensitivity, power levels, system life, expected failure rate, and cost.

A LED is a PN junction semiconductor that emits light when biased in the forward direction. Typically a current between 25 and 100 mA is switched through the diode, with the wavelength of the emission a function of the material used in doping the diode.

The laser diode offers users a fast response time in converting electrically encoded information into light encoded information; thereby, they can be used for very high data transfer rate applications. The laser diode can couple a high level of optical power into an optical fiber, resulting in a greater transmission distance than obtainable with light emitting diodes. Although they can transmit further at higher data rates than LEDs, laser diodes are more susceptible to temperature changes and their complex circuitry makes such devices more costly.

Optical cables

Many types of optical cables are marketed, ranging from simple 1-fiber cables to complex 18-fiber, commonly jacketed, cables. In addition, a large variety of specially constructed cables can be obtained on a manufactured-to-order basis from vendors.

In its most common form, an optical fiber cable consists of a core area, cladding, and a protective coating. This is illustrated in the upper part of Figure 7.2. As a light beam travels through the core material, the ratio of its speed in the core to the speed of light in free space is defined as the refractive index of the core.

A physical transmission property of light is that while traveling in a medium of a certain refractive index, if it should strike another material of a lower refractive index, the light will bend towards the material containing the higher index. Since the core material of an optical fiber has a higher index of refraction than the cladding material, this index differential causes the transmitted light signal to reflect off the core–cladding junctions and propagate through the core. This is shown in the lower part of Figure 7.2.

The core material of fiber cables can be constructed with either plastic or glass. While plastic is more durable to bending, glass provides a lower attenuation of the transmitted signal. In addition, glass has a greater bandwidth, permitting higher data transfer rates when that material is used for fiber construction.

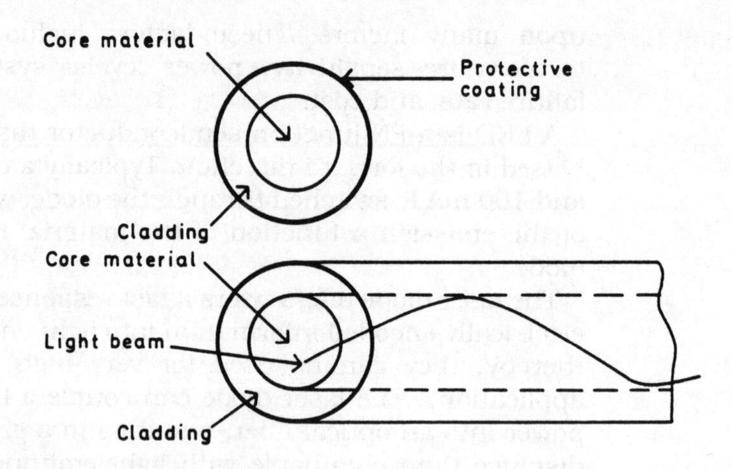

Figure 7.2 Optical fiber cable.

The capacity of a fiber optic data link of a given distance depends upon the numerical aperture (NA) of the cable as well as the core size, attenuation, and pulse dispersion characteristics of the fiber. The NA value can be computed from the core and cladding refractive indices as follows:

$$NA = \sqrt{m_1^2 - m_2^2}$$

where: m_1 = core material refractive index, and m_2 = cladding material refractive index.

The numerical aperture value indicates the potential efficiency in the coupling of the light source to the fiber cable. Together with core material diameter, the numerical aperture indicates the level of optical power that can be transmitted into a fiber.

A function of both the fiber material and core/cladding imperfections, attenuation determines how much power can reach the far end of an unspliced link. When the numerical aperture, core diameter, and attenuation are considered together you can determine the probable transmission power loss ratio.

Pulse dispersion is a measure of the widening of a light pulse as it travels down an optical fiber. Dispersion is a function of the cable's refractive index. Fibers with an appropriate refractive index permit an identifiable signal to reach the light detector at the far end of the cable.

Types of fibers

Two types of optical fibers are available for use in cables: step index and graded index. A step index optical fiber is also commonly referred to as a singlemode fiber optic cable, while a graded index fiber is also known as a multimode graded index fiber optic cable.

In a step index (singlemode) fiber an abrupt refractive index change exists between the core material and the cladding which results in lightwaves traveling in a relative straight line through the cable. In a (multimode) graded index fiber the refractive index varies from the center of the core to the core–cladding junction. The gradual variation of the refractive index in this type of fiber serves to minimize the optical signal dispersion as light travels along the fiber core. The minimization of dispersion results from the light rays near the core–cladding junction traveling faster than those near the center of the core. The minimization of signal dispersion reduces attenuation to less than 5 dB per kilometer, allowing transmission to occur over greater distances at higher data rates than available with a step index fiber. Thus, graded index fibers are commonly used for long-haul communications applications. Conversely, step index fibers are used for communications applications requiring a narrow bandwidth and relatively low data rate requirement; however, the absence of a slow bending or stepping effect results in a very low attenuation. This in turn enables very long transmission links with few repeaters.

Cable size

Fiber optic cables are represented by a pair of numbers separated from one another by a slash (/). The first number represents the outer diameter size of the core, while the second number represents the outer diameter size of the cladding. Both measurements are in microns or millionths of a meter.

Common cable types

Five common types of optical cables are illustrated in Figure 7.3. Note that when a 1-fiber cable has a transmitter connected to one cable end and a receiver to the opposite end, it normally functions as a simplex transmission medium. As such,

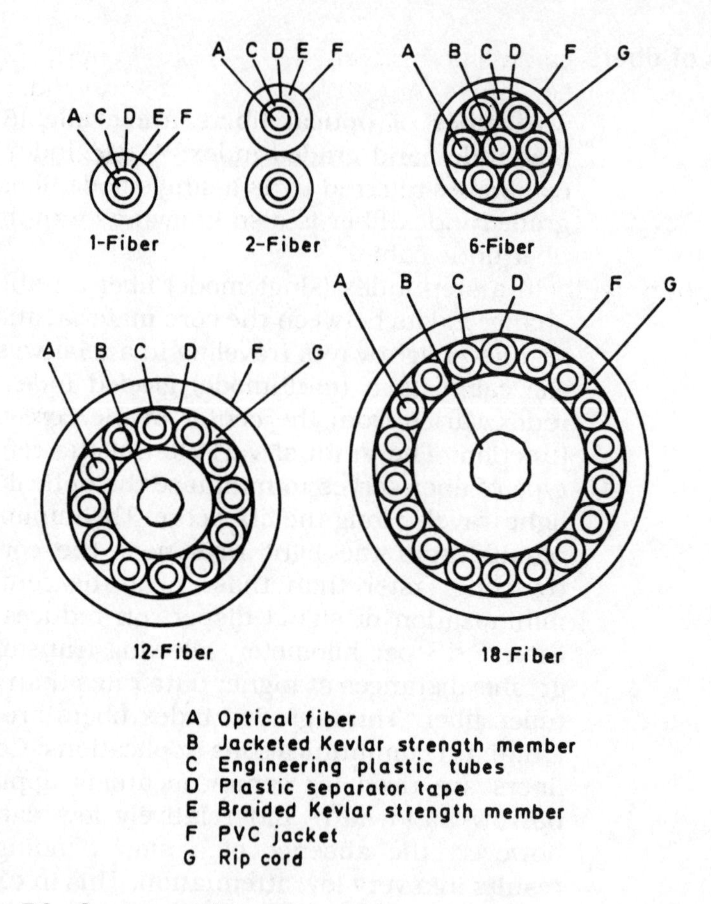

Figure 7.3 Common cable types.

transmission can only occur in one direction. The 2-fiber cable can be considered a duplex transmission medium since each fiber permits transmission to occur in one direction. When transmission occurs on both fibers at the same time a full-duplex transmission sequence results. Thus, the 18-fiber cable is capable of providing nine duplex transmission paths.

The exception to simplex transmission on a 1-fiber cable is obtained through the use of wavelength division multiplexing. This technique, which is based upon frequency division multiplexing, results in the utilization of two or more different light wavelengths to subdivide the bandwidth of a fiber cable into subchannels similar to the manner in which FDM divides a circuit into subchannels by frequency. Today, wavelength division multiplexing is commonly used by communications carriers to obtain a full-duplex transmission capability on long-distance fiber optic circuits. In addition, several communications equipment vendors have incorporated wavelength division multiplexing into fiber optic modems built into their fiber optic multiplexers, enabling full duplex transmission to be obtained over a 1-fiber cable.

The light detector

To convert the received light signal back into a corresponding electrical signal a photodetector and associated electronics are required. Photodetection devices currently available include a PIN photodiode, an avalanche photodiode, a phototransistor, and a photomultiplier. Due to their efficiency, cost, and light signal reception capabilities at red and near infrared (IR) wavelengths, PIN and avalanche photodiodes are most commonly employed as light detectors.

In comparison to PIN detectors, avalanche photodiode detectors offer greater receiver sensitivity. This increased light sensitivity results from their high signal-to-noise ratio, especially at high bit rates.

Since avalanche photodiode detectors require an auxiliary power supply which introduces noise, circuitry to limit such noise results in the device having a higher overall cost than a PIN photodiode. In addition, they are temperature sensitive and their installation environment requires careful examination. A block diagram of the major components of a light detector is illustrated in Figure 7.4.

7.2 THE OPTICAL MODEM

The transmission of information on a fiber optic cable is obtained by the pulsing of a light source. Since data terminal equipment, including computer ports and terminal devices, transmit information as a series of voltage pulses, a conversion device is required to translate electrical voltages into light pulses. This conversion device is an optical modem which can also be considered as an 'optical modulator' and 'optical demodulator,' since it impresses information onto a carrier (light) by varying the carrier (turning light on and off).

In its broadest sense an optical modem is a device housing both an optical transmitter and an optical detector as shown in the top portion of Figure 7.5. Similar to conventional modem

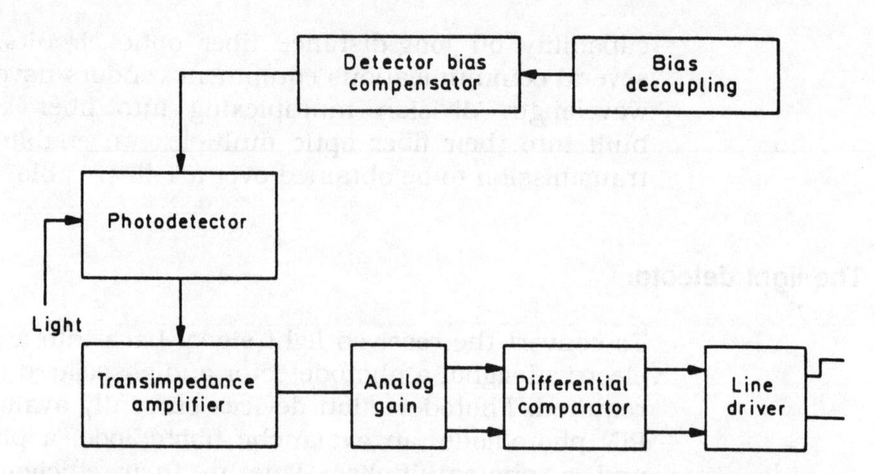

Figure 7.4 Light detector components. To filter DC input voltage, protect the photodiode, and reduce the effect of electromagnetic interference, a light detector module has bias decoupling. Since an avalance photodiode is a temperature sensitive device, a detector-bias compensator is used to compensate for temperature changes that could affect the diode.

The photodetector converts the received optical signal into a low-level electrical signal. This detector can either be an avalance or PIN diode, depending upon the optical sensitivity requirements. The transimpedence amplifier is a low-noise, current-to-voltage converter while the analog gain element increases the voltage gain from the preceding amplifier to the level required for the decision circuitry. The differential comparator converts the analog signal into digital form by interpreting signals below a certain threshold as a '0' and above that threshold as a '1'. The line driver regenerates and drives the squared signal from the comparator for transmission over metallic cable.

development, a variety of optical modems have evolved. These variations range from single-channel stand-alone devices to multiport optical modems, the latter capable of functioning as a synchronous multiplexer and optical modem. The multiport optical modem permits conventional electronic bit streams from up to four data sources to be multiplexed and converted into one stream of light pulses for transmission on one optical fiber. In the lower part of Figure 7.5, a multiport optical modem servicing four data sources is shown.

Both full- and half-duplex optical modems are commonly available. A half-duplex fiber optic modem interfaces a 1-fiber cable and does not perform wavelength division multiplexing. A full-duplex fiber optic modem either interfaces two fiber cables as illustrated in the top portion of Figure 7.5 or interfaces a 1-fiber cable and performs wavelength division multiplexing to obtain two separate transmission paths.

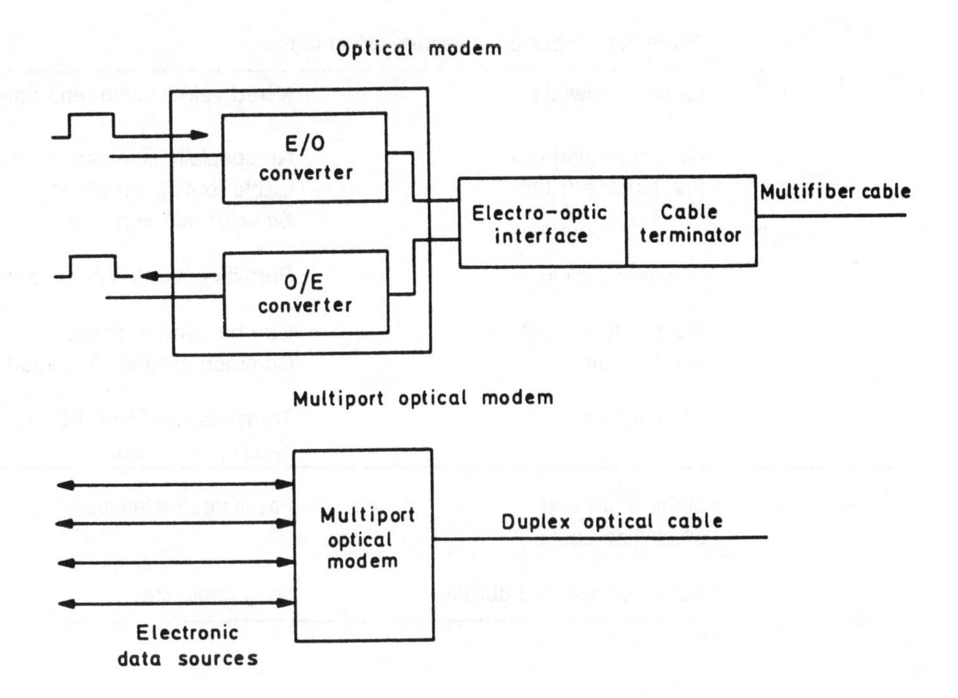

Figure 7.5 Optical modem. An optical modem can transmit and receive data over one multifiber cable, converting the electronic data source to light and the received light back into its corresponding electronic signal. Serving as a synchronous multiplexer, the multiport optical modem transmits data from up to four electronic sources as one optical signal.

7.3 OPTICAL TRANSMISSION ADVANTAGES AND LIMITATIONS

When used for data transmission, fiber optic cables offer many advantages over cables with metallic electrical conductors. These advantages result from several distinct properties of the optical cables. The more common advantages associated with the utilization of a fiber optic transmission system are listed in Table 7.1.

Bandwidth

One of the advantages of fiber optic cable in comparison to metallic conductors is the wide bandwidth of optical fibers. With potential information capacity directly proportional to transmission frequency, light transmission on fiber cable provides a transmission potential for very high data rates. Currently, data rates of up to 10^{14} bps have been achieved

Table 7.1 Fiber optic system advantages

Large bandwidth	Mixed voice, video, and data on one line
No electromagnetic interference (EMI)	No specially shielded conduits required Cable routing simplified Bit error rate improved
Low attenuation	Permits extended cable distances
No shock, hazard, or short circuits	Can be used in dangerous atmospheres Common ground eliminated
High security	Transmission TEMPEST acceptable Tapping noticeable
Lightweight and small-size cable	Facilitates installation
Cable rugged and durable	Long cable life

on fiber optic links. When compared to the data transmission limitation of telephone wire pairs, fiber cable makes possible the merging of voice, video, and data transmission on one conductor. In addition, the wide bandwidth of optical fiber provides an opportunity for the multiplexing of many channels of lower speed, but which are still significantly higher data rate channels than are transmitted on telephone systems.

Electromagnetic non-susceptibility

Since optical energy is not affected by electromagnetic radiation, optical fiber cables can operate in a noisy electrical environment. This means that special conduits, formerly required to shield metallic cables from radio interference produced by such sources as electronic motors and relays, are not necessary.

Similarly, cable routing is easier since the rerouting necessary for metallic cables around fluorescent ballasts does not cause concern when routing fiber cables. Due to its electromagnetic interference immunity, fiber optic transmission systems can be expected to have a lower bit error rate than corresponding metallic cable systems. In fact, you can expect data errors to be reduced by a factor of approximately 10 000 using optical fibers in comparison to coaxial cable or shielded twisted-pair. By not generating electromagnetic radiation, fiber optic cables do not generate crosstalk. This property permits multiple fibers to be routed in one common cable, simplifying the system design process.

Signal attenuation

The signal attenuation of optical fibers is relatively independent of frequency. In comparison, the signal attenuation of metallic cables increases with frequency. The lack of signal loss at frequencies up to 1 GHz permits fiber optic systems to be expanded as equipment is moved to new locations. In comparison, conventional metallic cable systems may require the insertion of line drivers or other equipment to regenerate signals at various locations along the cable.

Electrical hazard

On fiber optic systems light energy is used in place of electrical voltage or current for the transfer of information. The light energy alleviates the potential of a shock hazard or short circuit condition. The absence of a potential spark makes fiber optic transmission particularly well suited for such potentially dangerous industrial environment uses as petrochemical operations as well as refineries, chemicals plants, and even grain elevators. A more practical benefit of optical fibers for most corporate networks is the complete electrical isolation they afford between the transmitter and receiver. This results in the elimination of a common ground which is a requirement of metallic conductors. In addition, since no electrical energy is transmitted over the fiber, most building codes permit this type of cable to be installed without running the cable through a conduit. This can result in considerable savings when compared to the cost of installing a conduit required for conventional cables, whose cost can exceed $2500 for a 300 ft metal pipe.

Security

Concerning security, the absence of radiated signals makes the optical fiber transmission TEMPEST acceptable. In comparison, metallic cables must often be shielded to obtain an acceptable TEMPEST level. Although fibers can be tapped like metallic cable, doing so would produce a light signal loss. Such a loss could be used to indicate to users a potential fiber-tap condition.

Weight and size

Optical fibers are smaller and lighter than metallic cables of the same transmission capacity. As an example of the significant differences that can occur, consider an optical cable of 144 fibers with a capacity to carry approximately 100 000 telephone conversations. The cable would be approximately one inch in diameter and weigh about 6 ounces per foot. In comparison, the equivalent-capacity copper coaxial cable would be about three inches in diameter and weigh about 10 pounds per foot. Thus, fiber optic cables are normally easier to install than their equivalent metallic conductor cables.

Durability

Although commonly perceived as being weak, glass fibers have the same tensile strength as steel wire of the same diameter. In addition, cables containing optical fibers are reinforced with both a strengthening member inside and a protective jacket placed around the outside of the cable. This permits optical cables to be pulled through openings in walls, floors, and the like without fear of damage to the cable.

With better corrosion resistance than that of copper wire, transmission loss at splice locations has a low probability of occurring when optical fiber cable is used.

Limitations of use

As discussed, optical fiber cables offer many distinct advantages over conventional metallic cables. Unfortunately, they also have some distinct disadvantages. Two of the main limiting factors of fiber optic systems are cable splicing and system cost.

Cable splicing

When cable lengths of extended distances become necessary, optical fibers must be spliced together. To permit the transmission of a maximum amount of light between spliced fibers, precision alignment of each fiber end is required. This alignment is time consuming and, depending upon the method used to splice the fibers, the installation cost can rapidly escalate.

Fibers may be spliced by welding, gluing, or through the use of mechanical connectors. All three methods result in some degree of signal loss between spliced fibers.

Welding or the fusion of fibers results in the least loss of transmission between splice elements. The time required to clean each fiber end and then align and fuse the ends with an electric arc does not make splicing easily suitable for field operations. An epoxy, or gluing method of splicing requires the use of a bonding material that matches the refractive index of the core of the glass fiber. This method typically results in a higher loss than obtained with the welding process. While mechanical connection has the highest data loss among the three methods, it is by far the easiest method to employ. Although mechanical connectors considerably reduce splicing time requirements, the cost of good-quality connectors can be relatively expensive. Currently, typical connectors cost approximately $20.

System cost

Good quality, low-loss single-channel fiber optic cable costs between $1.50 and $2 per meter. A typical fiber optic modem having a transmission range of 1 km costs between $200 and $600. In comparison, conventional metallic cable for synchronous data transmission costs approximately 30 cents per foot, while a line driver capable of regenerating digital pulses at data rates up to 19.2 kbps cost approximately $200.

Based upon the preceding, the cost of a limited-distance fiber optic system at data rates up to 19.2 kbps will generally exceed an equivalent conventional metallic-based transmission system. Only when high data rates are required or transmission distances expand beyond the capability of line drivers and, to some extent, limited-distance modems, are fiber optic systems economically viable.

7.4 UTILIZATION ECONOMICS

One of the most commonly used duplex fiber optic cables cost \$2.50 per meter which is equivalent to 76 cents per foot. An optical modem containing an electric-to-optic and an optic-to-electric converter capable of transferring data at rates up to 10 Mbps costs approximately \$600 per unit. The system cost of a pair of optical modems as a function of distance is illustrated in Figure 7.6. Note that as long as no splicing is required, costs are a linear function of distance.

Suppose a requirement materializes which calls for the communications linkage of four high-speed digital devices located in one building of an industrial complex to a computer center located in a different building 10 000 ft away. What fiber optic systems can support this requirement and what are the economic ramifications of each configuration?

Dedicated cable system

Equivalent to individually connecting devices on metallic cables, four individual fiber optic systems and four cables could

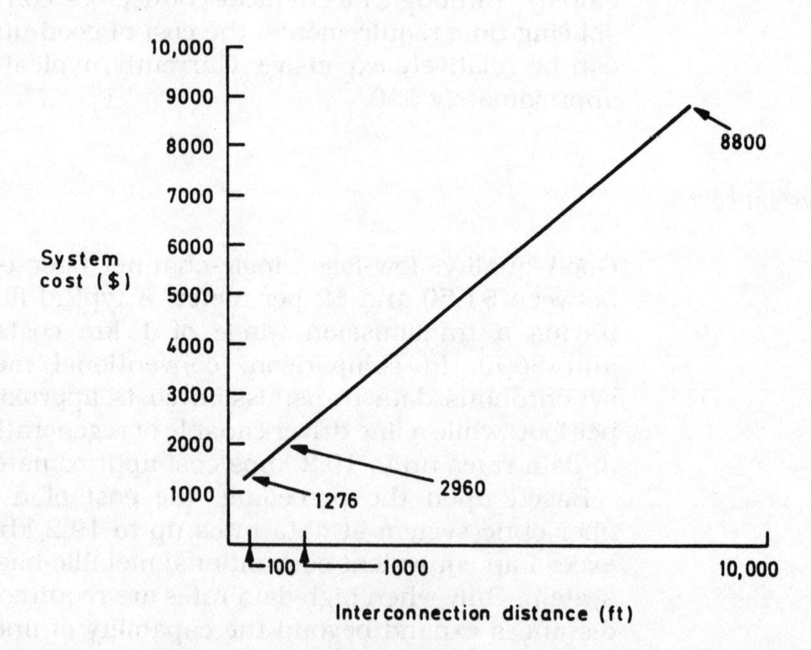

Figure 7.6 System cost varies with distance. Cost of a typical duplex optical fiber cable and a pair of optical modems capable of transmitting up to 10 Mbps.

be installed. Here eight optical modems would be required, resulting in the cost of the modems being $4800.

With four cables being required, 40 000 ft of cable would cost $30 400, resulting in a total cost of $35 200 for this type of network configuration. In addition, a substantial amount of personnel time may be required to install four individual 10 000 ft cables.

Multichannel cable

A second method that can be employed to link multiple devices at one location is by the employment of multichannel cable. We can examine the economics associated with multichannel cable by considering the cost of an eight-channel cable.

A typical eight-channel cable capable of supporting four duplex transmissions costs approximately $10 per meter or $3.05 per foot. On a duplex channel basis, this represents a cost of 76 cents per foot per channel. This is the same cost as an individual duplex cable. The use of most multichannel cable does not offer any appreciable savings over individual cable until ten or more channels are packaged together. Prior to excluding the use of a multichannel cable, you should carefully consider cable installation costs since the time required to install one multichannel cable can be significantly less than the time required to install several individual cables.

Optical multiplexers

Similar to conventional metallic cables, the potential installation of parallel optical cables indicates that multiplexing should be considered. Prior to deciding upon the use of an optical multiport modem or optical multiplexer, you should examine the type of data to be transmitted as well as the data transfer rates required. If each data type is synchronous and no more than four data sources exist, the utilization of an optical multiport modem can be considered. If a mixture of asynchronous and synchronous data must be supported or, if more than four data sources exist, you should consider an optical multiplexer.

Currently, four-channel synchronous optical multiport modems cost approximately $500 per unit. Returning to our system requirement, a pair of optical multiport modems and one cable could support the four data terminals. Here the total system cost would be reduced to $8600, of which $1000 would be for the pair of optical multiport modems and the remainder for the cable.

Suppose one or more of the data terminals was an asynchronous device, or more than four terminals required communications support. For such situations an optical multiplexer should be considered. One optical multiplexer currently marketed costs $1250 and supports up to eight data channels at data rates up to 64 kbps. Since an optical receiver/transmitter is included in the multiplexer, only the cost of one duplex cable must be added to the cost of a pair of multiplexers. Doing so, we will obtain the cost of an optical transmission system capable of supporting up to eight data sources at data rates up to 64 kbps. For such a system the cost of 10 000 feet of cable and the two optical multiplexers would be $9100. Although the cost of an optical multiplexer system slightly exceeds the cost of using of optical multiport modems, the use of the multiplexers provides the ability to expand support for four more devices without any additional cost.

In Table 7.2 the reader will find a comparison of the four previously discussed network configurations. As indicated, both cost and expansion capacity varies widely between configurations.

Table 7.2 Network configuration comparisons of four data sources and 10 000 ft transmission distance

Individual cable	Multichannel cable	Optical multiport modem	Optical multiplexer
System cost			
$35 200	$35 200	$8600	$9100
Expansion capability:			
Extra cable and transmitter/ receiver per data source	Requires more expensive cable and additional transmitters/ receivers	Cannot support more than four channels	No cost to add support for up to four additional channels

7.5 TYPES OF FIBER OPTIC MULTIPLEXERS

In a previous chapter we reviewed the advantages of statistical time division multiplexers in comparison to conventional TDMs. In spite of those advantages, all fiber optic multiplexers are designed to operate using time division multiplexing. Since fiber optic multiplexers in many instances represent the latest advances in multiplexing technology, its use of time division multiplexing may appear to be both strange and inefficient. In actuality, its use of conventional time division multiplexing is based upon the tremendous bandwidth obtainable from the transmission of light on fiber optic cable, which eliminates the necessity for seeking greater efficiency through the use of statistical multiplexing. For example, the bandwidth on a voice-grade analog line is approximately 3000 Hz, which limits the operating rate of modems used on this transmission facility to the range 19.2 to 24 kbps. This in turn limits the mixture of data sources a multiplexer can support when an analog transmission facility is used to interconnect multiplexers, which resulted in the development of statistical multiplexers. In comparison, a fiber optic cable may support a bandwidth in the MHz to GHz range, resulting in an increase in bandwidth up to one million times that obtainable on analog transmission facilities. This in turn enables the fiber optic multiplexer to operate without the channel capacity constraints associated with the use of TDMs and STDMs connected to conventional transmission facilities.

Most fiber optic multiplexers contain a built-in fiber optic modem. Thus, the operational capability of the multiplexer to a degree is dependent upon the built-in fiber optic modem, including the type of fiber connection it supports.

Selection considerations

Table 7.3 lists three key areas for fiber optic multiplexer selection consideration areas as well as the selection elements in each area readers may wish to evaluate.

Typical asynchronous channel support ranges in increments of 8 channels to 64 channels. Unlike conventional TDMs that do not pass control signals from an attached DTE, many fiber optic multiplexers can be obtained to support this feature. However, in supporting the passing of control signals the fiber optic multiplexer uses a separate channel. Thus, a fiber optic multiplexer that supports, for example, 8 asynchronous

Table 7.3 Fiber optic multiplexer selection considerations

Data source support:	Asynchronous channels
	Synchronous channels
	Mixed channels
DTE physical interface	RS-232
	Coaxial cable
	Twinaxial cable
	V.35
	T1
Fiber optic modem	Single/dual fiber support
	Wavelength division multiplexing support

ports, including selected control signals, would support 16 asynchronous ports if control signals are not passed through the multiplexer.

Unlike asynchronous control signals that require the use of an extra port if passed through a fiber optic multiplexer, synchronous control signals can be passed on the same port. To accomplish this, the synchronous data rate is internally increased by the multiplexer to allocate a time slice for the inclusion of samples of control signals. Similar to asynchronous channel support, most fiber optic multiplexers that support synchronous transmission do so in increments of 8 channels, with some equipment supporting up to 144 channels.

Data rates supported by synchronous channels are dependent upon the channel interface. Some fiber optic multiplexers are limited to supporting RS-232, which limits data rates to approximately 19.2 kbps. Other fiber optic multiplexers support V.35 and T1 interfaces, enabling data rates up to 64 kbps and 1.544/2.048 Mbps, respectively, depending upon whether the T1 interface is designed for North American (1.544 Mbps) or European (2.048 Mbps) data sources.

Although some fiber optic multiplexers require separate channels for asynchronous and synchronous data source support, other multiplexers include channels that support both transmission modes. Normally, channels that support both asynchronous and synchronous data sources are restricted to an RS-232 interface which limits the operating rate of the channel to 19.2 kbps.

Physical interface

The physical interface on each fiber optic multiplexer channel governs the maximum data rate support of the channel. Although RS-232 is normally limited to 19.2 kbps, some vendors can support up to 38.4 kbps. Similarly, the normal V.35 data rate of up to 64 kbps has been extended to 76.8 kbps, while T1 operating rates of 1.544 Mbps and 2.048 Mbps are commonly supported. The ability to support coaxial and twinaxial cable connections enables some fiber optic multiplexers to be used in application-specific areas. Such application-specific areas include functioning as an IBM 3X74 control unit extender or as an IBM System 3X extender.

Figure 7.7 illustrates the use of a fiber optic multiplexer as an IBM 3X74 control unit extender. In this example, the fiber optic multiplexer interfaces the coaxial cable connection of an IBM 3X74 control unit, enabling devices located thousands of feet away from the control unit to share the use of a common cable instead of 32 separate cable runs. In addition to saving on cabling cost, which can easily exceed the cost of a pair of fiber optic multiplexers and one fiber cable, this design also reduces the potential for cable congestion and the problems associated with the weight of coaxial cable. Not only is coaxial cable bulky, but, in addition, the weight of the cable bundled with other cables can cause problems if they should become entangled with conventional telephone wire. Thus, replacing 32 bulky and heavy coaxial cables with one fiber optic cable can also provide room for growth as well as eliminate potential cable entanglement problems.

Similar to the previously described control unit extenders, some vendors have tackled the problem of twinaxial cable

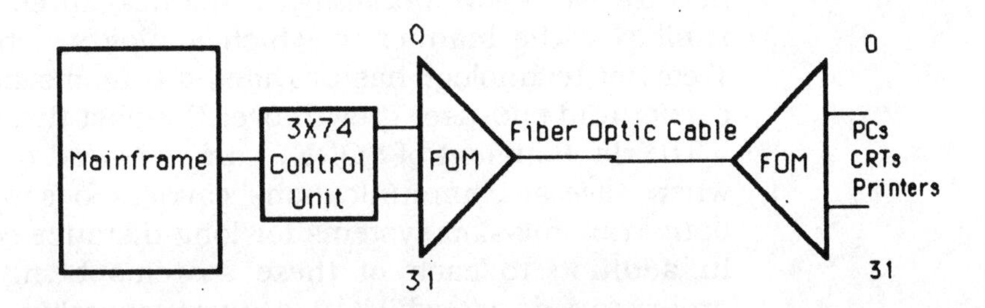

Figure 7.7 IBM 3X74 control unit extender. When used as a 3X74 control unit extender, the fiber optic multiplexer (FOM) supports the connection of coaxial cable at its physical interface.

used with IBM System 3X minicomputers. In doing so vendors manufactured fiber optic multiplexers that could be cabled to the twinaxial ports of minicomputers, enabling peripherals to be located remote from the computer via the use of a relatively economical fiber optic cable.

Fiber optic modem

The third and, for some potential fiber optic multiplexer users, most important area of consideration is the built-in fiber optic modem. The type of modem used in the fiber optic multiplexer will govern its support of a single fiber for full-duplex transmission or the requirement to use a multiple fiber cable to obtain this capability. In addition, the type of electric-to-optical converter and the support of a specific cable type and size will govern the transmission distance obtainable between multiplexers—another key constraint you must consider.

7.6 SONET AND SONET MULTIPLEXERS

In concluding this chapter on fiber optic multiplexers, we will examine both a transport vehicle capable of delivering data at gigabit rates as well as the optical multiplexers used to place data onto fiber optic cable carried by that transport vehicle. The transport vehicle we will examine is the Synchronous Optical Network (SONET) which provides the potential to radically alter communications for both carriers and users due to the operating rates it supports. Via the use of fiber optic highways, SONET can be expected to provide support for a new transmission hierarchy for the next three or four decades, similar to the manner in which analog wideband and digital T-carrier technology has provided a transmission hierarchy for carrier and end-user growth over the past three decades.

The development of SONET can be traced to the early 1980s when several communications carriers began to install fiber optic transmission systems for long-distance communications. In addition to each of these systems being based upon a proprietary design which made internetworking difficult at best, in common they lacked the ability to easily extract a single DS0 channel which represents one digitized voice signal at a 64 kbps operating rate. Recognizing these problems, Bellcore, the R&D laboratory of the Bell Operating Companies, developed SONET. SONET was proposed by Bellcore to the American National Standards Institute (ANSI) in 1985 and by ANSI to the CCITT in 1986. By 1989, ANSI and the CCITT had adopted the SONET transmission hierarchy which is listed in Table 7.4. In this table the term OC references Optical Carrier, which is an optical interface. Although SONET definitions exist for up to 255 levels, as indicated in Table 7.4, only eight levels are currently standardized.

Table 7.4 The SONET hierarchy

Level	Line rate (Mbps)
OC-1	51.84
OC-3	155.52
OC-9	466.56
OC-12	622.08
OC-18	933.12
OC-24	1244.16
OC-36	1866.24
OC-48	2488.32

Frame structure

The SONET frame, which is shown in Figure 7.8, consists of nine rows of 90 bytes, or 810 bytes, of which the first 3 bytes in each row are used as pointers. These pointers indicate where each DS0 channel in the SONET frame begins, which enables individual DS0s to be easily extracted and reinserted. In comparison, in the T1 hierarchy, a T3 signal must be first demultiplexed to T2 and then demultiplexed to T1 to extract a single DS0.

The SONET Synchronous Transport Signal level 1 (STS-1) frame illustrated in Figure 7.8 repeats at the 8000 frame per second rate of T1 and E1. Thus, the operating rate of STS-1 becomes 810 bytes × 8 bits/byte × 8000 samples/second, or 51.84 Mbps. Since 27 bytes in each frame are used as pointers, the data handling capacity of SONET's STS-1 becomes 50.112 Mbps.

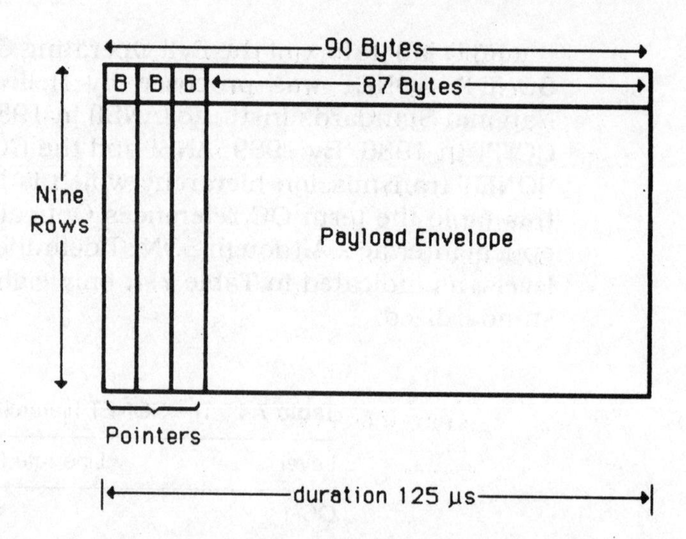

Figure 7.8 SONET frame structure signal level-1. A basic SONET frame contains 810 bytes, of which 783 can each carry one 64 kbps DS0 channel.

The payload envelope is referred to as a virtual container, as each byte represents a 64 kbps bit rate. Thus, each byte has a capacity equivalent to a single DS0. Since a T1 circuit contains 24 DS0s, while an E1 circuit consists of 32 DS0s, the STS-1 frame structure is capable of transporting 32 T1 or 24 E1 circuits.

By packing three STS-1 frames into a higher-level frame the OC-3 data transport rate becomes available. That is, three STS-1 frames when multiplexed result in a total of 2430 bytes per frame. This results in an operating rate of 2430 bytes × 8 bits/byte × 8000 samples/second, or 155.52 Mbps. Similarly, OC-9, OC-12, and higher SONET levels represent the multiplexing of 9, 12, and additional STS-1 frames.

Possible applications

Since SONET does not define an interface below 51.84 Mbps, for most organizations the use of this technology will appear to be far off into the future. However, as carriers begin to migrate to SONET it is a reasonable expectation that large organizations will require an entry to the STS-1 signal level. To provide this entry we can expect vendors to manufacture STS-1 frame equipment that will both enable access to public optical networks as well as provide a mechanism to use STS-1 as a data transmission highway between two organization locations.

Figure 7.9 illustrates how a SONET multiplexer might be used to carry the communications requirements of an organization via a single fiber optic cable to a carrier's central office. At that office the multiplexed data could be broken out into its individual services, such as T1, T3, fractional T1 (FT1), fractional T3 (FT3), and other services. Although the SONET multiplexer illustrated in Figure 7.9 is not currently available to end-users, within a few years we can expect its arrival.

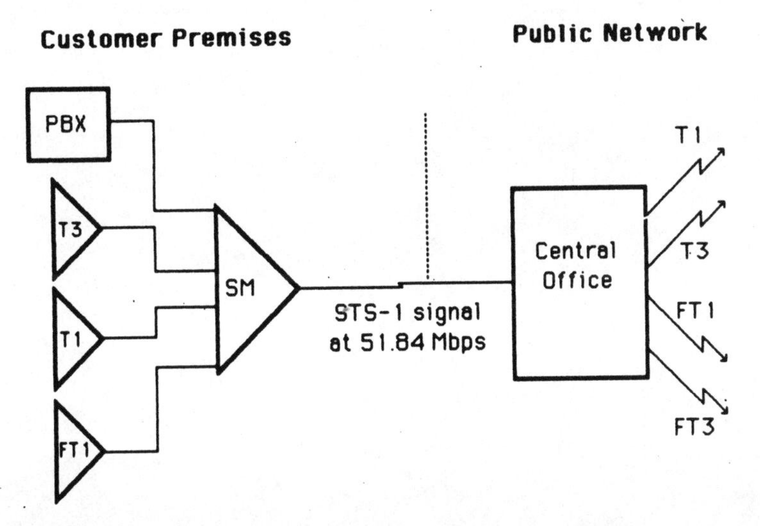

Figure 7.9 The SONET transportation highway. The use of a SONET multiplexer (SM) can be expected to provide organizations with the ability to use a common fiber optic cable from a carrier's central office for all of its communications requirements.

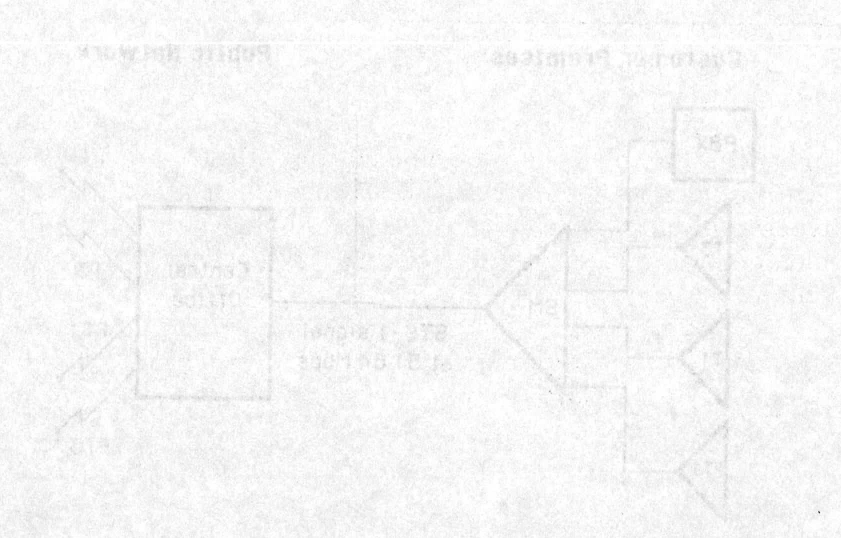

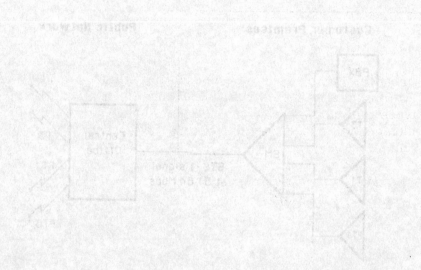

8

EVOLVING TECHNOLOGIES

In concluding this book, we will focus our attention upon a few evolving technologies whose use is applicable to two or more types of multiplexers. Technologies examined in this chapter include low bit-rate voice digitization, fast packet multiplexing, and frame relay.

8.1 LOW BIT-RATE VOICE DIGITIZATION

Low bit-rate voice digitization references any technique whereby a voice conversation is digitized at a rate under the 64 kbps PCM operating rate. Although a variety of voice digitization techniques are commonly available for use with T-carrier multiplexers, until recently they were not available for use with other types of multiplexers. This changed during 1991 as several vendors introduced TDMs and STDMs that are capable of supporting optional low bit-rate voice digitization modules.

Advantages

Through the use of low bit-rate voice digitization modules the ability to integrate voice and data has been extended to TDMs and STDMs which place digitized voice on a bandpass channel, precluding flow control causing voice delays. In addition to TDMs and STDMs, several vendors have made available low bit-rate voice digitization modules for their fractional T1 (FT1) multiplexers, which increases the capability of this category of communications equipment to integrate voice, data, and video.

Applications

Figure 8.1 illustrates one example of how an STDM supporting the use of low bit-rate voice digitization could be used to integrate voice and data. In this example it was assumed that a dual CVSDM (continuous voice slope delta modulation digitization module) module card digitizes each voice connection from a PBX at a data rate of 16 kbps, resulting in the STDM reserving 32 kbps from the available 56 or 64 kbps bandwidth for voice on a bandpass channel. Thus, the remaining 24 or 32 kbps bandwidth would be available for statistical multiplexing or 16 kbps from the remaining bandwidth could be used to support an additional digitized voice connection to the PBX.

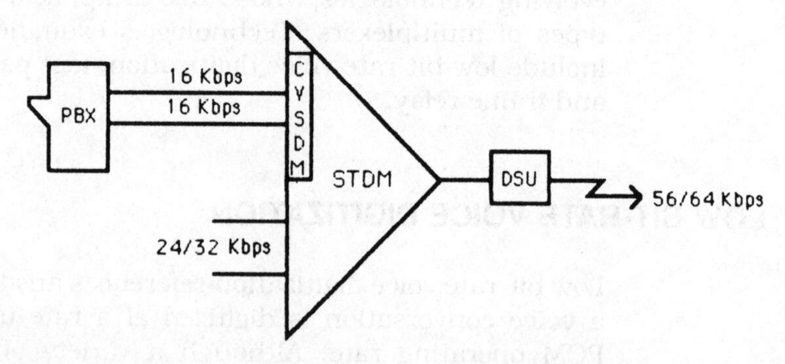

Figure 8.1 STDMs using low bit-rate voice digitization. Through the use of low bit-rate voice digitization modules, both voice and data can be carried on subrate transmission facilities.

In the area of fractional T1, several vendors originally introduced multiplexers which support FT1 service by accepting 64×*N* or 56×*N* data sources while producing an aggregate T1 or E1 line operating rate. Although these fractional T1 multiplexers resemble economy T1 multiplexers since they were designed to simply time division multiplex *N* data channels, their lack of voice digitization support limited their utilization to interfacing digital PBXs if they were to carry voice conversations. Recognizing this limitation, vendors have recently introduced FT1 multiplexing equipment which support optional voice digitization modules. This in turn has enabled network users to support voice and data on FT1 circuits without the necessity to obtain full-functional T-carrier multiplexers.

8.2 FAST PACKET MULTIPLEXING

Fast packet multiplexing can be considered as a generic term which references the ability of a multiplexer to dynamically transport different types of data sources, including voice, data, and video, in a packetized format. Originally developed for use with T-carrier multiplexers, fast packet multiplexing has been incorporated into communications products ranging in scope from multiplexers that interface 56/64 kbps transmission facilities to devices that can be considered as modified FT1 multiplexers. For each of these products, fast packet technology provides an operational capability similar to a statistical multiplexer. In fact, the primary difference between an STDM and a fast packet multiplexer is that the former builds frames that include data from many active inputs and whose frame integrity is achieved by the use of a CRC which provides for error detection and correction. While the use of a CRC per frame promotes data integrity, it also results in delays as frames flow through a network since error checking occurs at each node. This can result in unpredictable delays, which makes STDM unsuitable for carrying voice through an STDM network.

Operation

In a fast packet multiplexer, data sources are packetized and packets are prioritized for transportation. For example, digitized voice would have a higher priority than data, while synchronous data that could have a timeout if delayed would be serviced prior to asynchronous data. Concerning network delay, fast packet multiplexing examines the data source to determine if error detection and correction is being performed by communications equipment and/or an application. If so, fast packet multiplexing will not perform error detection and correction, but relies instead upon the equipment or application to provide data integrity. This in turn reduces packet delays as data is routed through a fast packet network.

Figure 8.2 illustrates the operation of a fast packet multiplexer. In fast packet multiplexing all packets are of the same length, use the same number of addressing bits, and are routed through a network using the same routing technique. Packets differ from one another only with respect to their content and destination. In the example illustrated in Figure 8.2 note that each active data source is assigned the maximum line

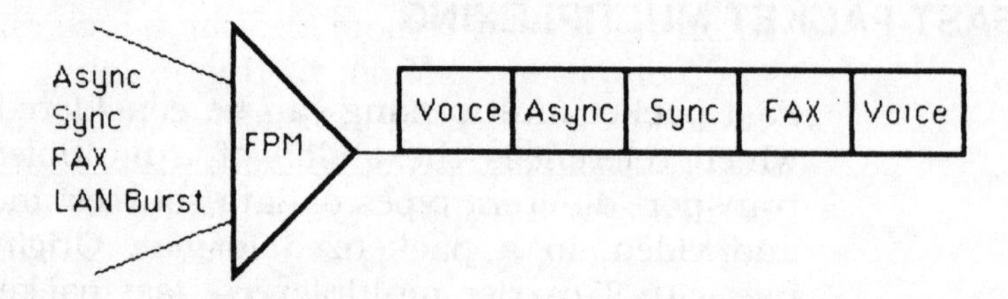

Figure 8.2 Fast packet multiplexing. All packets are of the same length, use the same number of addressing bits, and are routed through a network using the same routing technique.

bandwidth for filling one packet. Thus, a 2400-bps synchronous input would fill a packet flowing at 1.544 Mbps if the fast packet multiplexer was connected to a T1 circuit.

Applications

Since the full bandwidth is allocated to each active data source, the fast packet multiplexer is better suited for handling LAN traffic bursts and provides a better mechanism for interconnecting LANs than fixed operating rate channels that may be idle most of the time. The ability to prioritize packets enables voice to be transported without delay, enabling oversizing of the multiplexer with respect to a TDM-based 24- or 30-channel T1/E1 multiplexer to occur without worrying about voice delays.

Standards

Unlike frame relay which is a standards-based protocol and is discussed in Section 8.3, fast packet technology is a proprietary technology. Although your readings of trade publications and other books, including some titles by this author, make you well aware of the advantages of standards, the proprietary nature of fast packet technology by itself is not a major obstacle to its effective utilization. That is, organizations can use this technology as a transport mechanism to carry voice, data, and video rapidly and efficiently between remote locations where many data sources cannot tolerate delays. For example, fast packet multiplexers could be utilized in a manner similar to the configuration shown in Figure 8.3. In this example, a fast packet multiplexer is used to provide a transport mechanism to interconnect two remote locations, including their PBXs, LANs, and computer systems. In this example, you would configure each fast packet multiplexer to prioritize digital voice while you could let computer-to-computer and interLAN transmission contend for available bandwidth when voice conversations are active. Since LAN traffic is bursty it would be carried more effectively by fast packet technology that provides it with access to the full transmission bandwidth when LAN traffic is transferred. In comparison, conventional multiplexing would require interLAN traffic to be carried at a fixed data rate regardless of whether there was heavy or light interLAN traffic.

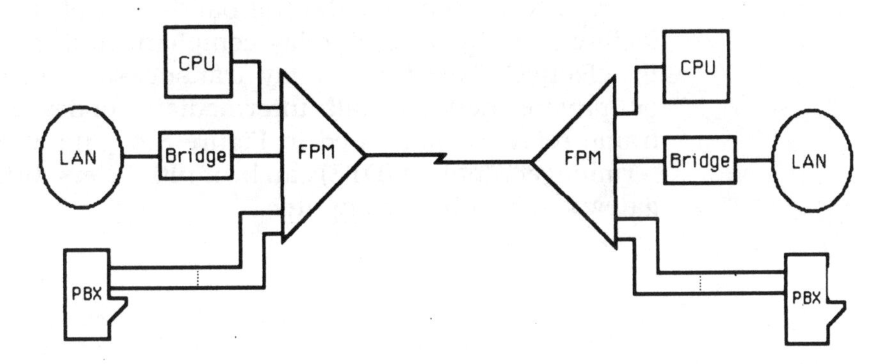

Figure 8.3 Using fast packet multiplexers. Fast packet multiplexers are well suited to handle bursty traffic from LANs as well as voice conversations.

8.3 FRAME RELAY

Frame relay is a standards-based protocol which is similar to X.25 packet switching but which eliminates transmission delays due to the absence of error checking at intermediate nodes. Unlike fast packet technology which provides a mechanism to prioritize packets and delay other packets, frame relay treats all packets as being similar in priority via a lack of an ability to interrupt selective packet transmission. Thus, fast packet technology is suitable for carrying voice, data, and video, while frame relay is restricted to carrying data.

Protocol support

In spite of the previously mentioned limitation, frame relay represents an excellent technology for providing data transportation for essentially all communications protocols. This is because frame relay encapsulates native mode information, such as SDLC, bisync, async, TCP/IP, and other protocols to form frame relay frames. Those frames are then passed end-to-end through a frame relay network and converted back into their original format at the destination by the removal of the encapsulated information.

Operation

Similar to fast packet technology, frame relay enables all data sources to contend for the full bandwidth of the transmission facility used by a frame relay compliant multiplexer. However, to effectively use frame relay transmission through a public or private network, all intermediate nodes must support frame relay as illustrated in Figure 8.4. The frame relay data terminal equipment (DTE) can be multiplexers, bridges, routers, gateways, or a host computer.

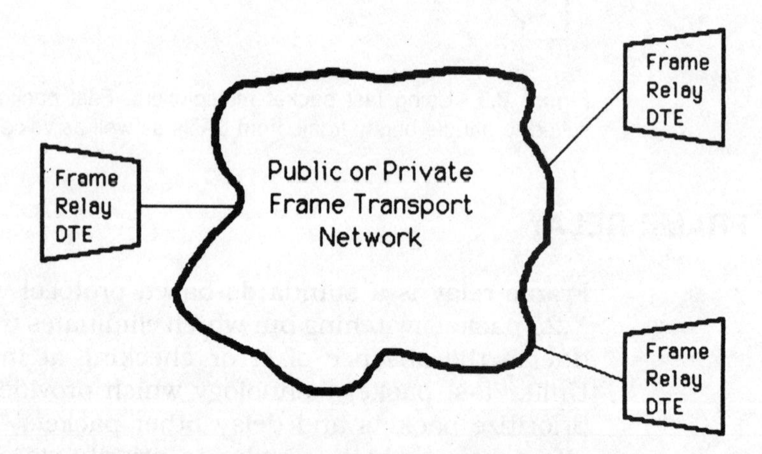

Figure 8.4 Networking frame relay DTEs. To use frame relay all intermediate nodes must support this technology.

8.4 ALTERNATIVES TO CONSIDER

The major rationale and driving force behind both fast packet and frame relay technology is the ability of these technologies to use bandwidth more efficiently. For example, a circuit switching

T-carrier multiplexer requires each data source to be serviced by one or more fixed-speed channels regardless of the actual activity on the channel. In comparison, both fast packet and frame relay dynamically assign available bandwidth based upon activity, enabling bursts of activity to be handled more efficiently.

You can obtain a similar capability to the dynamic bandwidth allocation capability of fast packet and frame relay technology via the use of STDMs by overspeeding STDM channels. That is, by servicing data sources at double or triple their conventional operating rate, you can let the STDM service bursty traffic more efficiently. Of course, when too many inputs are active, the STDM will implement flow control and become the arbitrator of which data source acquires the use of the available bandwidth of the transmission facility for a small period of time. However, this is no different than the situation where too many frame relay or fast packet data sources become active and a frame relay or fast packet multiplexer must then arbitrate access to bandwidth. However, since STDMs implement error checking at iNtermediate nodes, the data flow through an STDM network will result in a delay exceeding the delay associated with implementing a frame relay or fast packet network. In spite of the additional delay, overspeeding STDM inputs can provide an extension to the life of that equipment and delay the necessity to move to newer technology.

Figure 8.5 illustrates an example of STDM overspeeding. In this example, a 32-port IBM 3174 control unit with a token-ring adapter connected to a 60-workstation LAN that might normally operate at 19.2 kbps is converted to a wideband interface and clocked into the STDM at 56 kbps. A remote job entry terminal which would normally operate at 9.6 kbps is clocked into the STDM at 19.2 kbps, while a Hewlett Packard 3000 computer performing distributed system (DS) file transfer operations is also serviced at 19.2 kbps instead of its nominal rate of 9.6 kbps. Lastly, a ten-position rotary connected to ten 9.6 kbps asynchronous modems is serviced by ten 19.2 kbps ports. This configuration results in the STDM accepting 286.4 kbps input and arbitrating input traffic onto a 64 kbps link.

The example of overspeeding an STDM represents an actual network configuration which the author can discuss from experience. Since 3174 terminals and LAN workstations are interactive in nature, their communications requirements seldom interfere with RJE or DS operations that are periodic in nature. In fact, from IBM's Network Performance Monitor

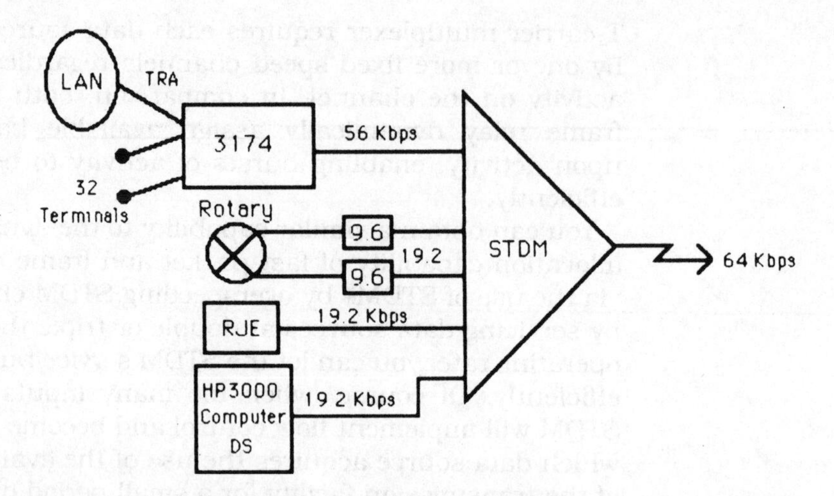

Figure 8.5 Overspeeding STDMs. By overspeeding inputs into an STDM, the bandwidth can be used more effectively for bursty traffic.

statistics which queries the 3174 to obtain response time statistics it was determined that 80% of all transactions had a response time under one second. In comparison, prior to overspeeding the 3174, its operation at 19.2 kbps resulted in only 40% of all transactions having a response time under one second.

The key to the use of STDM overspeeding is the fact that the network using this technology is based upon a star topology and although multi-node STDMs are used at the central site, only two hops are required to route data between any network location. If the number of hops were to increase, a corresponding increase in response time would occur due to error checkIng at intermediate nodes. Thus, for networking applications in which the number of hops is limited, overspeeding STDMs may provide an alternative to the performance associated with fast packet and frame relay technology.

INDEX

Access request operation 106–8
ACCUNET T1.5 118
Adaptive differential pulse code modulation (ADPCM) 135
Alternate routing 7
Ancillary device control 110–11
ASCII code 28, 44, 113
Asynchronous channel adapters 39
AT&T Teletypewriter Exchange Service (TWX) 28
Auto code detection 90–1
Auto echo 91
Auto speed detection 90–1

Bandpass feature 92
Bandpass multiplexing 85–6
Binary 8 zero substitution (B8ZS) 126
Binary speed 112
Break signal procedure 111
Buffer control 79–83
Buffer delay 81–3
Buffer occupancy 88
Buffer pair 40
Buffer utilization 88

Carriage return
  line feed insertion after 112
  padding after 111
Carrier telegraph systems 26–8
CCITT 22
CCITT FDM standard group 23–4
CCITT FDM standard mastergroup 25–6
CCITT FDM standard supergroup 24–6
CCITT FDM subchannel allocations 31
CCITT FDM wideband recommendations 23
CCITT G. recommendation 23
CCITT Higher Level Data Link Control (HDLC) frame structure 76
CCITT R.31 recommendation 27
  frequency assignments 27
CCITT X series recommendations 100–3
CCITT X.3 parameters 108–14
CCITT X.3 recommendation 102, 106, 108–14
CCITT X.25 recommendation 100–1
  data or link level 101
  packet level 101
  physical level 101
CCITT X.28 commands 114–16
CCITT X.28 recommendation 102
CCITT X.29 recommendation 102–3
Cell relay 19
CEPT PCM-30 system 120
Channel banks 118–19
  D1 118
  D-type 117, 118
  versus T-carrier multiplexers 120
Channel service units (CSUs) 48
Character delete 113
Character error rate ratio 90
Circuit switched network 9
Clear channel DSO 126
Clear to send (CTS) control signal 14, 79
Clocking signal adjustment 79–80
Codec 118
Combined FDM–TDM 58–61
Command console 92
Communications carrier systems 22–6
Compression efficiency ratio 89
Consultative Committee for International Telephone and Telegraph *See* CCITT
Continuous variable slope delta modulation (CVSDM) 135–8
Cyclic redundancy check (CRC) 75, 78

D-type channel banks 117, 118
D1 channel banks 118
Data circuit-terminating equipment 100
Data communications equipment (DCE) 66, 67
Data compression 92
Data forwarding signal 110
Data service units (DSUs) 40, 48, 56
  see also Split stream DSUS
Data terminal equipment (DTE) 67, 100
Dataphone Digital Service (DDS) digital transmission facilities 70
Diagnostic testing 5

Differential PCM (DPCM) 135
Digital access cross connect (DAC)
  operations 139–41
Digital channel bank *See* D-type channel bank
Discard output 111

E1 carrier 118
E1 circuits 17, 117
E1 multiplexers 9, 17, 117
E1 system 120
Echo 110
Echo mask 113–14
Editing 113
Electrical hazard 163
Electromagnetic interference 162–3
E&M signaling 134–5
ENQ–ACK sequence 80
Error density 89
Error detection and correction 6

Fast packet multiplexers 179–81
  applications 180
  operation 179–80
  standards 180–1
Fiber optic modem 169, 172
Fiber optic multiplexers 18–19, 153–75
  absence of electrical hazard 163
  advantages of 161–4
  bandwidth 161–2
  cable routing 162–3
  cable size 157
  cable splicing 165
  cable types 157–9
  cable weight and size 164
  data rate support 170, 171
  dedicated cable system 166–7
  durability 164
  economics 166–8
  fiber types 157
  light detector 159
  light source 154–5
  limitations of use 164–5
  multichannel cable 167
  optical cables 155–9
  optical model 159–60
  physical interface 171
  security 163–4
  selection considerations 169–70
  signal attenuation 163
  system components 154–9
  system cost 165
  TEMPEST 163–4
  time division multiplexing 169
  types of 169–72
Flow control techniques 79–80, 98, 112
  variations to 80–1
Flyback delay 92–3
Frame relay 19, 181–2
  alternatives to consider 182–4
  data terminal equipment (DTE) 182
  operation 182
  protocol support 181–2
Frames transmitted 88–9
Frequency division multiplexers (FDM) 10–12, 21–35
  channel spacings 30
  data multiplexers 26–35
  data multiplexing capability 30–3
  FM operation 29
  full-duplex transmission mode 32–3
  types of 21–2
  utilization 33–5
  versus TDM 37–8
  see also Combined FDM–TDM
Frequency shift keying (FSK) 11, 28
Front-end substitution 50–1

Graded index fibers 157

Half-duplex method of transmission 72
High-speed modems (HSM) 57
Hub-bypass multiplexing 49–50

IBM 3X74 control unit extender 171
Idle timer delay 110
Inband signaling 79
Intelligent time division multiplexers (ITDM) 86–7
Inverse multiplexing 51–2
I/O channel adapter 38–40

Laser diode (LD) 154
Light emitting diode (LED) 18, 154
Line delete 113
Line display 113
Line driver 119
Line feed
  insertion after carriage return 112
  padding after 112–13
Line folding 111
Local area networks (LANs) 19
Logical channel number (LCN) 104
Loop signaling 133–4
Low bit-rate voice digitization 177–8
  advantages 177
  applications 178
Low-speed modems (LSM) 56

MegaStream 119
Mesh network 151–2
Multidrop lines 98
Multiplexer centers 52
Multiplexer loading 88
Multiplexers
  economics 2–4
  evolution and utilization 10–19
  future developments 19
  rationale for using 2–9

Multiplexing
  alternative configurations 56–8
  cost reductions 52–61
  economics 52–61
  networking example 53–5
Multipoint network 149–51
Multiport modems 61–8
  application example 63–6
  channel combinations 64
  data communications equipment (DCE) interface 67
  daytime/night-time operations 66
  error detector 67
  operation 61–2
  selection criteria 62–3
  selector switches 66
  standard and optional features 66–8
  test pattern generator 67

Negative acknowledgements (NAKs) 88–9
Network configurations, comparisons of data sources 168
Network management support 6–7
Network structures 146–52
  see also under specific network structures
Nodal processors 145
Null character padding 111
Number of frames transmitted 88–9

Open System Interconnection (OSI) reference model 101
Optical cables 155–9, 157
  parallel 167
Optical fibers 157
Optical modem 159–60
Optical multiplexers 167–8
  comparison of data sources 168
  economics 168
Optical multiport modem 167
Optical transmission *See* Fiber optic multiplexers; Optical multiplexers
Outband signaling 79

Packet assembler/disassemblers (PADs) 15–17, 99–116
  access request operation 106–8
  command state 105–6
  data transfer state 106
  evolution 99–100
  flow control 112
  operation 104–5
  packet assembly process 104
  packet framing 104–5
  recall 110
  service signal control 111
  states 105
  terminal identifier 107–8
  X.28 command signals 114–16
Packet multiplexers, alternatives to consider 182–4
Packet switched network 9
Padding
  after carriage return 111
  after line feed 112–13
Page wait 114
Parity treatment 114
Permanent terminal 108
Phantom channel 10
Point-to-point multiple link network 146
Point-to-point multiplexers 47–8, 56
Point-to-point single link network 146
Port contention 93, 95–6
Port group support 96
Pulse code modulation (PCM) 17, 118

Repeaters 119
Ring network 149
Routing capability 8

Series multipoint multiplexing 48
Service ratios 83–7
SONET and SONET multiplexers 19, 172–5
  applications 174–5
  development 172–3
  frame structure 173
  hierarchy 173
  payload envelope 174
  Synchronous Transport Signal level 1 (STS-1) 173–4
  transportation highway 175
Split stream DSUs 68–70
  applications 69–70
  modes of operation 70
  networking capability 70
  stand-alone 69
Star network 147–9
Statistical loading ratio 89–90
Statistical multiplexers 13–15, 71–98
  address and byte count method 75–6
  applications 94–8
  buffer control 79–83
  buffer delay 81–3
  comparison to TDMs 71–9
  data integrity 78
  data source support 85–6
  dual addressing format 76–8
  efficiency 72, 84
  features of 90–4
  flow control 183
  flow control methods 81
  message frame construction 74–8
  multi-node feature 93
  multi-node networking 94–7
  networking capability 97–8
  operational problems 78–9
  overspeeding 183–4

Statistical multiplexers (*cont.*)
performance level 97–8
performance measurement information 89–90
protocol support 93
service ratios 83–7
statistics 87–90
statistics display 94
switching capability 96–7
topological configurations 94
Statistical time division multiplexers (STDMs) *See* Statistical multiplexers
STDM *See* Statistical multiplexers
Step index fibers 157
Stratacom fast packet multiplexer 132
Switching multiplexers 7–9, 96–7
Synchronization character 45–6
Synchronous channel adapters 40
Synchronous Optical Network *See* SONET and SONET multiplexers
Syntax error 74

T1 carrier 118
T1 circuit, evolution and operation 117–24
T1 circuits 17, 118
T1 multiplexers 117
relationship with CSU 126
T1 span line 119
T1/E1 multiplexers 9, 17
T-carrier multiplexers 17, 98, 117–52
ADPCM 135–6
alternate routing and route generation 142–3
application overview 127–8
bandwidth allocation 131–3
bandwidth utilization 129–30
cabinet layout 144–5
channel rates 128
comparison with nodal processor 145
configuration rules 144–5
CSU function 125–7
CVSDM 136–8
diagnostics 144
digital access cross connect capability 139–41
E&M signaling 134–5
efficiency 128
gateway operation support 141–2
internodal trunk support 138–9
loop signaling 133–4
major features of 129
maximum number of hops and nodes supported 144
multi-node 146
networking 146–52
operational characteristics 125–8
PCM 135
redundancy 143–4
subrate channel utilization 139
versus channel banks 120
voice digitization support 135–8
voice interface support 133–5
Telenet 108
Teletype Corporation Model 33 28
Teletypewriter Exchange Service (TWX) 28
TEMPEST 163–4
Terminal identifier 107–8
Time assigned speed interpolation (TASI) 135
Time division multiplexers (TDM) 11–14, 37–70, 119
applications 46–52
bit interleaving 43–5
central logic 40
character interleaving 43–5
comparison to STDMs 71
composite adapter 40–1
efficiency 45–6
frame format 46
front-end substitution 50–1
hub-bypass multiplexing 49–50
inverse multiplexing 51–2
message frame 41
message frame construction 72–4
message train 42
multiplexing and demultiplexing process 41
multiplexing interval 42–3
operation 38–46
point-to-point multiplexing 47–8
series multipoint multiplexing 48
techniques 43–5
versus FDM 37–8
see also Combined FDM–TDM; Multiport modems; Split stream DSUs
Transit delay 98
Tymnet PAD 106–8
Type B telegraph carrier system 26

Virtual container 174
Voice/data integration 9
Voice-grade lines 11, 51–2

XON/XOFF characters 110–11, 112
XON/XOFF sequence 79, 80

Index compiled by Geoffrey C. Jones